X

and the **City**

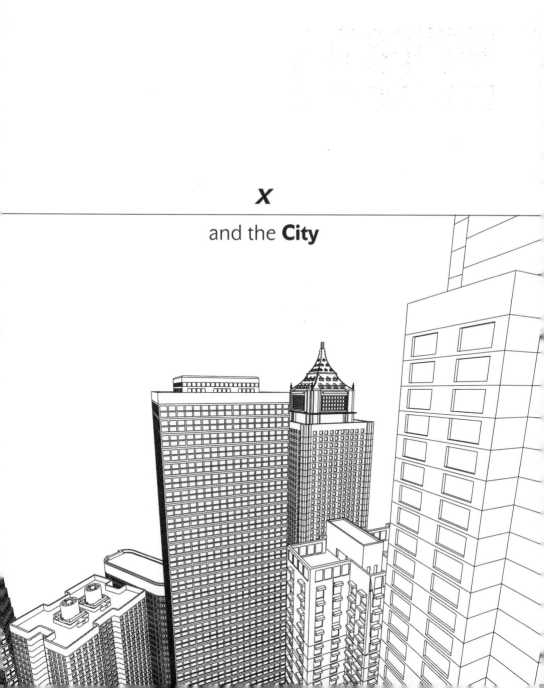

X and the City

MODELING ASPECTS OF URBAN LIFE

John A. Adam

PRINCETON UNIVERSITY PRESS / PRINCETON AND OXFORD

Library of Congress Cataloging-in-Publication Data
Adam, John A.
X and the city : modeling aspects of urban life / John Adam.
p. cm.
Includes bibliographical references and index.
ISBN 978-0-691-15464-0 1. Mathematical models. 2. City and town life—
Mathematical models. 3. Cities and towns—Mathematical models. I. Title.
HT151 .A288 2012
307.7601'5118—dc23   2012006113

British Library Cataloging-in-Publication Data is available

This book has been composed in Garamond

Book design by Marcella Engel Roberts

Printed on acid-free paper. ∞

Printed in the United States of America

1 3 5 7 9 10 8 6 4 2

For Matthew, who drifted "continentally" to a large city

He found out that the city was as wide as it was long and it was as high as it was wide. It was as long as a man could walk in fifty days . . . In the middle of the street of the city and on either bank of the river grew the tree of life, bearing twelve fruits, a different kind for each month. The leaves of the tree were for the healing of the nations.

—St. John of Patmos

(See Chapters 1 and 5 for some estimation questions inspired by these passages.)

CONTENTS

PREFACE

After the publication of *A Mathematical Nature Walk*, my editor, Vickie Kearn, suggested I think about writing *A Mathematical City Walk*. My first reaction was somewhat negative, as I am a "country boy" at heart, and have always been more interested in modeling natural patterns in the world around us than man-made ones. Nevertheless, the idea grew on me, especially since I realized that many of my favorite nature topics, such as rainbows and ice crystal halos, can have (under the right circumstances) very different manifestations in the city. Why would this be? Without wishing to give the game away too early into the book, it has to do with the differences between nearly parallel "rays" of light from the sun, and divergent rays of light from nearby light sources at night, of which more anon. But I didn't want to describe this and the rest of the material in terms of a city *walk*; instead I chose to couch things with an "in the city" motif, and this allowed me to touch on a rather wide variety of topics that would have otherwise been excluded. (There are *seven* chapters having to do with traffic in one way or another!)

As a student, I lived in a large city—London—and enjoyed it well enough, though we should try to identify what is meant by the word "city." Several related dictionary definitions can be found, but they vary depending on the country in which one lives. For the purposes of this book, a city is a large, permanent settlement of people, with the infrastructure that is necessary to make that possible. Of course, the terms "large" and "permanent" are relative, and therefore we may reasonably include towns as well as cities and add the phrase "or developing" to "permanent" in the above definition. In the Introduction we will endeavor to expand somewhat on this definition from a historical perspective.

This book is an eclectic collection of topics ranging across city-related material, from day-to-day living in a city, traveling in a city by rail, bus, and car (the latter two with their concomitant traffic flow problems), population growth in cities, pollution and its consequences, to unusual night time optical effects in the presence of artificial sources of light, among many other topics. Our cities may be on the coast or in the heartland of the country, or on another continent, but presumably always located on planet Earth. Inevitably, some of the topics are multivalued; not everything discussed here is unique to the city—after all, people eat, garden, and travel in the country as well!

Why *X and the City*? In the popular culture, the letter X (or x) is an archetype of mathematical problem solving: "Find x." The X in the book title is used to introduce the topic in each subsection; thus "$X = t_c$" and "$X = N_{tot}$" refer, respectively, to a specific length of time and a total population, thereby succinctly introducing the mathematical topics that follow. One of the joys of studying and applying mathematics (and finding x), regardless of level, is the fact that the deeper one goes into a topic, the more avenues one finds to go down. I have found this to be no less the case in researching and writing this book. There were many twists and turns along the way, and naturally I made choices of topics to include and exclude. Another author would in all certainty have made different choices. Ten years ago (or ten years from now), the same would probably be true for me, and there would be other city-related applications of mathematics in this book.

Mathematics is a language, and an exceedingly beautiful one, and the applications of that language are vast and extensive. However, pure mathematics and applied mathematics are very different in both structure and purpose, and this is even more true when it comes to that subset of applied mathematics known as *mathematical modeling* (of which more below). I love the beauty and elegance in mathematics, but it is not *always* possible to find it outside the "pure" realm. It should be emphasized that the subjects are complementary and certainly not in opposition, despite some who might hold that opinion. I heard of one mathematician who referred to applied mathematics as "mere engineering"; this should be contrasted with the view of the late Sir James Lighthill, one of the foremost British applied mathematicians of the twentieth century. He wrote, somewhat tongue-in-cheek, that pure mathematics was a very important part of applied mathematics!

Applied mathematics is often elegant, to be sure, and when done well it is invariably useful. I hope that the types of problem considered in this book can be both fun and "applied." And while some of the chapters in the middle of the book might be described as "traffic engineering," it is the case that mathematics is the basis for all types of quantitative thought, whether theoretical or applied. For those who prefer a more rigorous approach, I have also included Chapter 17, entitled "The axiomatic city." In that chapter, some of the exercises require proofs of certain statements, though I have intentionally avoided referring to the latter as "theorems."

The subtitle of this book is *Modeling Aspects of Urban Life*. It is therefore reasonable to ask: what *is* (mathematical) modeling? Fundamentally, mathematical modeling is the formulation in mathematical terms of the assumptions (and their logical consequences) believed to underlie a particular "real world" problem. The aim is the practical application of mathematics to help unravel the underlying mechanisms involved in, for example, industrial, economic, physical, and biological or other systems and processes. The fundamental steps necessary in developing a mathematical model are threefold: (i) to formulate the problem in mathematical terms (using whatever appropriate simplifying assumptions may be necessary); (ii) to solve the problem thus posed, or at least extract sufficient information from it; and finally (iii) to interpret the solution in the context of the original problem. This may include validation of the model by testing both its consistency with known data and its predictive capability.

At its heart, then, this book is about just that: mathematical modeling, from "applied" arithmetic to linear (and occasionally nonlinear) ordinary differential equations. As a little more of a challenge, there are a few partial differential equations thrown in for good measure. Nevertheless, the vast majority of the material is accessible to anyone with a background up to and including basic calculus. I hope that the reader will enjoy the interplay between estimation, discrete and continuum modeling, probability, Newtonian mechanics, mathematical physics (diffusion, scattering of light), geometric optics, projective and three-dimensional geometry, and quite a bit more.

Many of the topics in the book are posed in the form of questions. I have tried to make it as self-contained as possible, and this is the reason there are several Appendices. They comprise a compendium of unusual results perhaps (in some cases) difficult to find elsewhere. Some amplify or extend material discussed in the main body of the book; others are indirectly related, but nevertheless connected to the underlying theme. There are also exercises scattered throughout; they are for the interested reader to flex his or her calculus muscles by verifying or extending results stated in the text. The combination of so many topics provides many opportunities for mathematical modeling at different levels of complexity and sophistication. Sometimes several complementary levels of description are possible when developing a mathematical model; in particular this is readily illustrated by the different types of traffic flow model presented in Chapters 8 through 13.

In writing this book I have studied many articles both online and in the literature. Notes identifying the authors of these articles, denoted by numbers in square brackets in the text, can be found in the references. A more general set of useful citations is also provided.

ACKNOWLEDGMENTS

Thanks to the following for permissions:

Achim Christopher (Figures 23.1 and 23.10)

Christian Fenn (Figure 22.5)

Skip Moen (Figures 3.4 and 21.1)

Martin Lowson and Jan Mattsson, for email conversations about their work cited here.

Larry Weinstein for valuable feedback on parts of the manuscript.

Alexander Haußmann for very helpful comments on Chapter 22.

Bonita Williams-Chambers for help with Figure 10.2 and Table 15.1.

My thanks go to Kathleen Cioffi, who oversaw the whole process in an efficient and timely manner. I am most appreciative of the excellent work done by the artist, Shane Kelley, who took the less-than-clear figure files I submitted and made silk purses out of sows' ears! Many thanks also to the book designer, Marcella Engels Roberts, for finding the illustrations for the chapter openers and designing the book. Any remaining errors of labeling (or of any other type, for that matter) are of course my own.

I thank my department Chair, J. Mark Dorrepaal, for arranging my teaching schedule so that this book could be written in a timely fashion (and my graduate students could still be advised!).

As always, I would like to express my gratitude to my editor, Vickie Kearn. Her unhurried yet efficient style of "author management" s(m)oothes ruffled feathers and encourages the temporarily crestfallen writer. She has great insight into what I try to write, and how to do it better, and her advice is always invaluable. And I hope she enjoys the story about my grandfather!

Finally, I want to thank my family for their constant support and encouragement, and without whom this book might have been finished a lot earlier. But it wouldn't have been nearly as much fun to write!

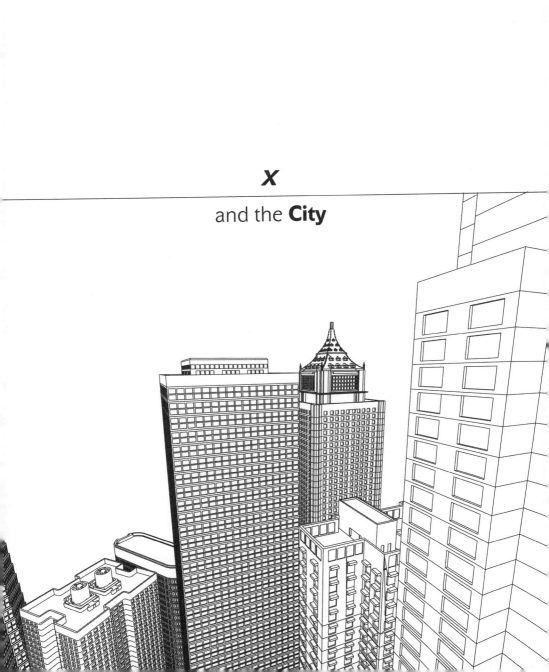

X

and the **City**

Chapter 1

INTRODUCTION
Cancer, Princess Dido, and the city

To look at the cross-section of any plan of a big city is to look at something like the section of a fibrous tumor.

—*Frank Lloyd Wright*

$X = ?$: WHAT ARE CITIES?

Although this question was briefly addressed in the Preface, it should be noted that the answer really depends on whom you ask and when you asked the question. Perhaps ten or twelve thousand years ago, when human society changed from a nomadic to a more settled, agriculturally based form, cities started to develop, centered on the Euphrates and Tigris Rivers in ancient Babylon. It can be argued that two hundred years ago, or even less, "planned" cities were constructed with predominately aesthetic reasons—architecture—in mind.

Perhaps it was believed that form precedes (and determines) function; nevertheless, in the twentieth century more and more emphasis was placed on economic structure and organizational efficiency. A precursor to these ideas was published in 1889 as a book entitled *City Planning, According to Artistic Principles*, written by Camillo Sitte (it has since been reprinted). A further example of this approach from a historical and geographical perspective, much nearer our own time, is Helen Rosenau's *The Ideal City: Its Architectural Evolution in Europe* (1983). But there is a distinction to be made between those which grow "naturally" (or organically) and those which are "artificial" (or planned). These are not mutually exclusive categories in practice, of course, and many cities and towns have features of both. Nevertheless there are significant differences in the way such cities grow and develop: differences in rates of growth and scale. Naturally growing cities have a slower rate of development than planned cities, and tend to be composed of smaller-scale units as opposed to the larger scale envisioned by city planners.

"Organic" towns, in plan form, resemble cell growth, spider webs, and tree-like forms, depending on the landscape, main transportation routes, and centers of activity. Their geometry tends to be irregular, in contrast to the straight "Roman road" and Cartesian block structure and circular arcs incorporated in so many planned cities, from Babylonian times to the present [1]. Some of the material in this book utilizes these simple geometric ideas, and as such, represents only the simplest of city models, by way of analogies and even metaphors.

ANALOGIES AND METAPHORS

Was Frank Lloyd Wright correct—do city plans often look like tumor cross sections writ large? Perhaps so, but the purpose of that quote was to inform the reader of a common feature in modeling. Mathematical models usually (if not always) approach the topic of interest using idealizations, but also sometimes using analogies and metaphors. The models discussed in this book are no exception. Although cities and the transportation networks within them (e.g., rail lines, roads, bus routes) are rarely laid out in a precise geometric grid-like fashion, such models can be valuable. The directions in which a city expands are determined to a great extent by the surrounding topography—rivers, mountains, cliffs, and coastlines are typically hindrances to urban growth. Cities are not circular, with radially symmetric population distributions, but even such gross idealizations have merit. The use of analogies in the mathematical sciences is well established [2], though by definition they have their limitations. Examples include Rutherford's analogy between the hydrogen atom and the solar system, blood flow in an artery being likened to the flow of water in a pipe, and the related (and often criticized) hydraulic analogy to illustrate Ohm's law in an electric circuit.

Analogy is often used to help provide insight by comparing an unknown subject to one that is more familiar. It can also show a relationship between pairs of things, and can help us to think intuitively about a problem. The opening quote by Frank Lloyd Wright is such an example (though it could be argued that it is more of a simile than an analogy). One possibly disturbing analogy is that put forward by W.M. Hern in the anthropological literature [3], suggesting that urban growth resembles that of malignant neoplasms. A neoplasm is an abnormal mass of tissue, and in particular can be identified with a malignant tumor (though this need not be the case). To quote from the abstract of the article,

> Malignant neoplasms have at least four major characteristics: rapid, uncontrolled growth; invasion and destruction of adjacent normal tissues (ecosystems); metastasis (distant colonization); and de-differentiation. Many urban forms are almost identical in general appearance, a characteristic that would qualify as "de-differentiation." Large urban settlements display "rapid, uncontrolled growth" expanding in population and area occupied at rates of from 5 to 13% per year.

The term "de-differentiation" means the regression of a specialized cell or tissue to a simpler unspecialized form. There *is* an interesting mathematical link that connects such malignancies with city growth—the *fractal dimension*. This topic will be mentioned in Chapter 18, and more details will be found in Appendix 9. For now, a few aspects of this analogy will be noted. The degree of border irregularity of a malignant melanoma, for example, is generally much higher than that for a benign lesion (and it carries over to the cellular level also). This is an important clinical feature in the diagnosis of such lesions, and it is perhaps not surprising that city "boundaries" and skylines are also highly irregular (in the sense that their fractal dimension is between one and two). This of course is not to suggest that the city is a "cancer" (though some might disagree), but it does often possess the four characteristics mentioned above for malignant neoplasms. The question to be answered is whether this analogy is useful, and in what sense. We shall not return to this question, interesting though it is, instead we will end this chapter by examining a decidedly non-fractal city boundary!

X = L: A CITY PERIMETER

In Greek mythology, Dido was a Phoenician princess, sister of Pygmalion, King of Tyre, and founder of the city of Carthage in northern Africa in 814 B.C. According to tradition, she did this in a rather unusual way. Pygmalion had her husband, Sychaeus, killed, and Dido fled to the northern coast of Africa. According to some, her brother agreed to let her have as much land as she could enclose within the hide of a bull; according to others, she bartered with the locals to accomplish this. She then cut the hide into a series of thin strips, joined them together, and formed a semicircular arc, with the Mediterranean Sea effectively as a diameter, thus enclosing a semicircular area [4]. Bravo! As we shall note in Appendix 1, given a straight boundary (as assumed in this story), a semicircle will contain the maximum area for a given perimeter, so Dido achieved the best possible result!

I grew up the son of a farm-laborer and had the occasional less-than-pleasant encounter with a certain bull; although he was British he was certainly not a gentleman. Here's how the mathematics might have gone. Imagine the typical bull torso to be a rectangular box 5 ft long by 4 ft high by 2 ft wide. (Yes, that's correct, we've cut off his legs, head, and tail, but only in our imagination.) The surface area of our bull-box is readily shown to be about 80 sq ft. We'll round this up to 100 sq ft since we're only interested in a rough estimate. And anyway, given how shrewd Dido appeared to be, no doubt she would have picked the biggest bull she could find! I don't know what kind of precision Dido or her servants might have had with the cutting tools available but I'm going to assume that strips could be cut as narrow as one hundredth of a foot (0.12 in, or about 3 mm). This may be an underestimate, but it makes the arithmetic easier without changing the "guesstimate" by much. If the total length of the strips is L ft, then we have a simple equation for the area: $0.01L \approx 100$, or $L \approx 10^4$ ft. This length would comprise the semicircular part of the city boundary. For a radius r ft,

$$L = \pi r \text{ and } A = \frac{1}{2}\pi r^2,$$

where A is the area of the "city." Hence

$$A = \frac{1}{2}\pi\left(\frac{L}{\pi}\right)^2 = \frac{L^2}{2\pi} \approx \frac{10^8}{2\pi} \text{ sq. ft} = \frac{10^8}{18\pi} \text{ sq. yd}$$
$$\approx \frac{10^8}{60 \times 3 \times 10^6} \text{ sq. mi} \approx 0.6 \text{ sq. mi.}$$

This is about 1.6 km². That's pretty impressive for a load of bull! If the boundary were circular instead of semicircular, the corresponding area would be half this amount, or about one third of a square mile. In Appendix 2 we shall generalize this idea to the case in which the boundary is of variable length $l < L$.

$X = V$: A CITY SIZE AND VOLUME

Let's return to part of the quotation from John of Patmos (stated at the beginning of this book). He was in exile on the island of Patmos, off the west coast of Turkey, probably around 90–95 A.D. Recall that, according to one translation, "He found out that the city was as wide as it was long and it was as high as it was wide. It was as long as a man could walk in fifty days." Another translation says that "it was twelve thousand furlongs in each direction, for its length, breadth and height are all equal." A furlong is 220 yards, there being eight in a mile, so the city was a cube 1,500 miles on a side. Perhaps with windows and doors and open spaces it was like a Menger sponge! (See Appendix 9.)

Question: How far might an adult walk in those days without the benefit of modern transportation?

I want to estimate this one! At 3 mph for 10 hours, 30 miles a day for 50 days is, guess what, 1,500 miles!

Question: What was the volume of the heavenly city in John's vision? In cubic miles? In cubic kilometers? In cubic furlongs? Estimate how many people it could accommodate.

An estimation question that can be generalized from the rest of the quotation—the number of leaves on a tree—can be found in Chapter 5.

Chapter 2

GETTING TO THE CITY

$X = y$: BY PLANE

Before we can explore mathematics in the city, we need to get there. The fastest way is to fly, so let's hop on a plane. If we assume that the descent flight path occurs in the same vertical plane throughout (i.e., no circling before touchdown—which is very rare these days), then the path is quite well represented by a suitably chosen cubic function [5], as we will see below. Here are some further conditions we impose prior to setting up a mathematical model for the path:

(i) The horizontal airspeed $dx/dt = U$ is constant throughout the descent. This is somewhat unrealistic, but we'll work with it anyway.

(ii) The descent begins when the plane is at the point $(-L, h)$, the origin being the beginning of the runway.

(iii) The magnitude of the vertical component of acceleration must not exceed the value k, where $0 \leq k \ll g$, $g = 32.2$ ft/sec^2 being the acceleration due to gravity.

(iv) The plane is considered for simplicity to be a point mass. The ultimate in cramped seating!

The landing path is modeled by

$$y(x) = ax^3 + bx^2 + cx + d. \tag{2.1}$$

What kind of boundary conditions should be imposed? Well, we would like to have a smooth touchdown, which means that not only do the wheels of the plane touch the runway at landing, they should have no downward component of velocity as they do so. That's the kind of landing that generates applause from the passengers! And at the beginning of the descent we would like to move smoothly down toward the airport, so a similar set of conditions applies, namely, that

$$y(0) = 0; \ y'(0) = 0;$$
$$y(-L) = h; \ y'(-L) = 0.$$

Exercise: Show that these conditions imply that the equation of the flight path is

$$y = h\left[2\left(\tfrac{x}{L}\right)^3 + 3\left(\tfrac{x}{L}\right)^2\right]. \tag{2.2}$$

A graph of $Y = y/h$ vs. $d = x/L$ is sketched in Figure 2.1.

The vertical component of the airplane's velocity is

$$u_y = \frac{dy}{dt} = \frac{dy}{dx}\frac{dx}{dt} = \frac{6Uh}{L}\left[\left(\tfrac{x}{L}\right)^2 + \left(\tfrac{x}{L}\right)\right],$$

and the corresponding acceleration component is

$$a_y = \frac{d^2y}{dt^2} = \frac{6U^2h}{L^2}\left[2\left(\tfrac{x}{L}\right) + 1\right].$$

Since a_y is an increasing function of x, its extreme values are $\pm 6U^2h/L^2$, occurring at the endpoints of the descent path (the minimum at the start and

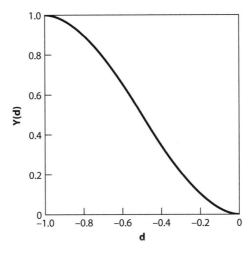

Figure 2.1. The flight path $\mathbf{Y(d)}$.

the maximum at the finish of the descent). Therefore from condition (iii) above,

$$\frac{6U^2h}{L^2} \le k. \tag{2.3}$$

This result can provide useful information depending on the type of flight and runway size. For transcontinental or transatlantic flights aboard a 747 jet (let's suppose the latter), we rewrite the above inequality as

$$L \ge \left(\frac{6U^2h}{k}\right)^{1/2}. \tag{2.4}$$

Typically, both U and h are "large" and k is "small," so this implies that L must be relatively large (compared with smaller airplanes). Suppose that $U = 600$ mph, $h = 40{,}000$ ft, and $L = 150$ miles. Then a further rearrangement of (2.4) gives a lower bound for k of approximately (in ft/s^2)

$$k_{\min} = 6\left(\frac{U}{L}\right)^2 h = 6 \times \left(\frac{600 \text{ miles}}{\text{hr}} \times \frac{1 \text{ hr}}{3600 \text{ s}} \times \frac{1}{150 \text{ miles}}\right)^2 \times 4 \times 10^4 \text{ ft}$$

$$= 6 \times \left(\frac{2}{9}\right)^2 \text{ ft/s}^2 = \frac{8}{27} \text{ ft/s}^2 \approx 0.30 \text{ ft/s}^2.$$

If, on the other hand, we find ourselves on a "puddle jumper" flight, then

$$U \leq \left(\frac{kL^2}{6h}\right)^{1/2}.$$

Therefore we expect L and k to be smaller than in the example, while h is still relatively large, so not surprisingly the airspeed will be correspondingly lower. Nevertheless, there are exceptions. Barshinger (1992) [5] relates his experience landing at Lake Tahoe. As the plane crossed the last peak of the 11000 ft Sierra Nevada Mountains, the airport was only twenty miles away. For these values of h and L and an airspeed of 175 mph, $k \approx 0.39$ ft/s², 30% larger than on the transatlantic flight!

Question: Can a similar analysis be carried out when equation (2.1) is replaced by $y = A + B\tanh C(x - D)$ (where $A, B, C,$ and D are constants)?

Well, I landed safely, and after checking in to my hotel I decided to take a walk, map in hand. Being directionally challenged at the best of times, even with a map, pretty soon I was lost, and it didn't seem to help much, because every place I wanted to go seemed to be right at the edge of the map, or on the well-worn fold! Getting irritated, I had a question to ask of no one in particular ...

$X = A$: THE MAP

Why is the place I'm looking for on a map *so often* at its boundary? Let's investigate. Consider first a map consisting of a rectangular sheet of relative dimensions 1 and $b < 1$, including a margin of width $a < b$. The area of the sheet is b square units and the relative area A of the margin is

$$A = \frac{\text{area of margin}}{\text{total area}} = \frac{2a}{b}(1 + b - 2a).$$

For a square map with unit side, $A(a) = 4a(1 - a)$. A has a maximum at $a = 0.5$; the map is all margin! More realistically, if $a = 0.1$, $A = 0.36$ or 36%, more than a third of the whole area is margin! Furthermore, the margin is 9/16 or about 56% of the map area!

Given a road atlas of the same relative dimensions, for the two pages combined,

$$A = \frac{2a}{b}\left(1 + 2b - 4a\right).$$

Taking $b = 1$ and $a = 0.1$ as before, $A = 0.52$, less than half the sheet is map!

X = $$: BY TAXI

Without finding my destination using the map, I decided to hail a taxi, but it seemed as if we drove forever. I'm sure we passed some places twice. Perhaps the driver took me on a world trip, which leads me to the following question . . .

Question: How many taxi rides would it take to circle the Earth?

Perhaps only one, if the driver is willing . . . more generally, we need to estimate the average length of a taxi ride and the circumference of the Earth. Let's start with the Earth. We might remember the circumference. If not, we might remember that the radius of the Earth is about 4000 miles and then calculate the circumference using $C = 2\pi r$. On the other hand, if we know that it is about 3000 miles from New York to Los Angeles and that they are three time zones apart, then 24 time zones will give you a circumference of 24,000 miles. This does seem a little more obscure, though.

Now let's consider a typical taxi ride. To go from downtown to Manhattan's Upper East (or West) Side is about 80 blocks or four miles [6]. Alternatively, the shortest ride will be about a mile and the longest will be about ten miles, so we can take the geometric mean[1] and estimate three miles. At three miles per ride, you will need about 8,000 taxi rides to circumnavigate the globe. Hey, that's about one per squirrel! (This is a teaser to keep you reading; see Chapter 6.) At $2.50 for the first 1/5 mile and $0.40 for each additional 1/5 mile, that

---

[1] This, by the way, is using the *Goldilocks principle*—is it too large, too small, or just right? The geometric mean helps us find "just right," often to within a factor of two or three, which is very useful when dealing with several orders of magnitude. Under these circumstances it is far more valuable than the arithmetic mean, though in this particular example the two give essentially the same answer. We shall encounter the need for this useful principle time and again in this book.

will cost you $20,000 plus $45,000 = $65,000 in cab fare (plus tips and waiting time). By the way: good luck hailing a cab in the middle of the Atlantic.

Since we'll be talking about driving in the city, and either flying or driving to get there, let's also apply these principles in that context and do some elementary "risk analysis."

Question: What is the risk (in the U.S.) of dying per mile traveled in a car? [7]

Okay: how far do Americans drive in a year? Warranties for 5 yr/60,000 miles are very common these days, so it's clear that car dealerships reckon that a typical domestic vehicle is driven about 12,000 miles per year on average. Alternatively, one can count the miles covered per week, halve it, and multiply by 100 (why?) Now there are a little over 300 million (3×10^8) people in the U.S., about half of whom drive (and maybe only half of those drive well), so the total mileage per year is about $1.5 \times 10^8 \times 1.2 \times 10^4 \approx 2 \times 10^{12}$ miles. As is readily checked, there are about 4×10^4 deaths due to car accidents in the U.S. each year, so the risk of death per mile is $4 \times 10^4 \div 2 \times 10^{12} = 2 \times 10^{-8}$ deaths/mile!

Alternatively, *what is the probability of being killed in a car accident?* The average life span of people in the U.S. is about 75 years, neglecting gender differences, so 1 in 75 Americans dies every year. Hence the average number of deaths per year is $3 \times 10^8 \div 75$, that is, 4×10^6 deaths/yr, so the total (lifetime) probability of dying in a car crash is 4×10^4 deaths/yr $\div 4 \times 10^6$ deaths/yr $= 0.01$, or 1%. It is therefore pretty safe to say that if you are reading this book, and are not a babe-in-arms, then your "probability" is rather less than this!

Question: What is the risk of dying per mile while flying? [7]

Most of us travel by air once a year (2 flights) on vacation or business, and a small fraction of the population travel much more than that. We'll use an average of 3 flights/person, so that's 10^9 people-flights per year; the actual figure in 2005 was 6.6×10^8, so we'll use the slightly more accurate value of 7×10^8 flights/yr. The average (intracontinental) flight distance probably exceeds 300 miles (or it would be simpler to drive) but is less than 3000 miles; the geometric mean is 10^3 miles, so we travel about 7×10^{11} mi/yr by air.

The crash frequency of large planes is (fortunately) less than one per year and (unfortunately) more than one per decade, so the geometric mean gives about one third per year. Typically, about 100 to 200 people die in each (large) crash; we'll take about 50 deaths/yr, so the per-mile probability of dying is, on the basis of these crude estimates, about $50 \div 7 \times 10^{11} \approx 7 \times 10^{-11}$ deaths/ mile, some 300 times safer than driving! Of course, this is overly simplistic since most crashes occur at take-off or landing, but it does remind us how safe air travel really is.

While on this subject, I recall hearing of a conversation about the risks of flying that went something like this:

"You know, the chances of dying on a flight are really very small. You're much more likely to die on the roads! And anyway, when it's your time to go, it's your time to go!'

"But what if it's the pilot's time to go and not mine?"

Chapter 3

LIVING IN THE CITY

$X = N$: A "ONE-FITS-MANY" CITY "BALLPARK" ESTIMATE

Suppose the city population is N_p million, and we wish to estimate N, the number of facilities, (dental offices, gas stations, restaurants, movie theaters, places of worship, etc.) in a city of that size. Furthermore, suppose that the average "rate per person" (visits per year, or per week, depending on context) is R, and that the facility is open on average H hours/day and caters for an average of C customers per hour. We shall also suppose there are D days per year. This may seem a little surprising at first sight: surely *everyone* knows that $D = 365$ when the year is not a leap year! But since we are only "guesstimating" here, it is convenient to take D to be 300. Note that 400 would work just as well—remember that we are not concerned here about being a "mere" factor of two or three out in our estimate.

Then the following ultimately dimensionless expression forms the basis for our specific calculations:

$$N \approx N_p \text{ people} \times \frac{R}{\text{year}} \times \frac{1 \text{ year}}{D \text{ days}} \times \frac{1 \text{ hour}}{C \text{ people}} \times \frac{1 \text{ day}}{H \text{ hours}}.$$

For simplicity, let's consider a city population of one million. How many dental offices might there be? Most people who visit the dentist regularly do so twice a year, some visit irregularly or not at all, so we shall take $R = 1$, $C = 5$, $H = 8$ and $D = 300$ (most such offices are not open at weekends) to obtain

$$N \approx 10^6 \text{ people} \times \frac{1}{\text{year}} \times \frac{1 \text{ year}}{300 \text{ days}} \times \frac{1 \text{ hour}}{5 \text{ people}} \times \frac{1 \text{ day}}{8 \text{ hours}} = \frac{10^6}{1.2 \times 10^4} \approx 100.$$

That is, to the nearest order of magnitude, about a hundred offices. A similar estimate would apply to doctors' offices.

Let's now do this for restaurants and fast-food establishments. Many people eat out every working day, some only once per week, and of course, some not at all. We'll use $R = 2$ per week (100 per year), but feel free to replace my numbers with yours at any time. The size of the establishment will vary, naturally, and a nice leisurely dinner will take longer than a lunchtime hamburger at a local "McWhatsit's" fast-food chain, so I'll pick $C = 50$, allowing for the fast turn-around time at the latter. Hours of operation vary from pretty much all day and night to perhaps just a few hours in the evening; I'll set an average of $H = 10$. Combining everything as before, with D equal to 400 now (such establishments are definitely open on weekends!),we find for the same size city

$$N \approx 10^6 \text{ people} \times \frac{100}{\text{year}} \times \frac{1 \text{ year}}{400 \text{ days}} \times \frac{1 \text{ hour}}{50 \text{ people}} \times \frac{1 \text{ day}}{10 \text{ hours}} = \frac{10^8}{2 \times 10^5} \approx 500.$$

Therefore the most we can say is that there are probably several hundred places to eat out in this city! I'm getting hungry . . .

We can do this for the number of gas stations, movie theaters, and any other facility you wish to estimate.

Exercise: Make up your own examples. Do your answers make sense?

Our final example will be to estimate the number of houses of worship in the city. Although many, if not most, have midweek meetings in addition to the main one at the end of the week or at some time during the weekend, I shall use the figure for R of 50 per year (or once per week) as above in the "eating out" problem. Spiritual food for those that seek it! But now I shall include the proportion of people who attend houses of worship in the calculation because clearly not everyone does. In the U.S. this is probably a higher proportion than in Europe, for example, so I shall suggest that one in five attend once per week in the U.S. The estimates for C and H are irrelevant (and meaningless) in this context, since everyone who attends regularly knows when the services start! Furthermore, this is more of a "discrete" problem since the vast majority of those who attend a house of worship do so once a week, so we shall simplify the formula by estimating an average attendance for the service. From my own experience, some churches have a very small attendance and some are "mega-churches," and I will assume that the range is similar for other faith traditions also. Using the Goldilocks principle—too small, too large, or just right?—I shall take the geometric mean of small attendance (10) and large attendance (1000), that is, 100. Hence the approximate number of houses of worship in a city of one million people is

$$N \approx 10^6 \, \text{people} \times \frac{1}{5} \times \frac{1}{100 \, \text{people}} = \frac{10^6}{5 \times 10^2} \approx 2000.$$

$X = d$: THE MUSEUM

Question: What is the optimal distance from which to view a painting/ sculpture/display?

At the outset, it should be pointed out that we are referring to an object for which the lowest point is above the eye level of the observer: a painting high on a wall, a sculpture or statue on a plinth, and so forth. Commonsense indicates that unless this is the case, the angle subtended by the statue (say) at the observer's eye will increase as the statue is approached. Of course, there is still an optimal viewing distance—wherever the observer feels most comfortable standing or sitting—but this is subjective. What this question means is where is the maximum angle subtended when the base or bottom of the painting is above the ground? That a maximum must occur is again obvious—far away that angle is very small, and close up it is also small, so a maximum must occur somewhere between those positions.

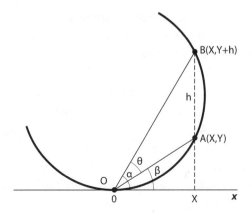

Figure 3.1. Display geometry for the maximum viewing angle. The object (e.g., a painting) lies along the vertical segment **AB**.

This is posed as a standard optimization problem in many calculus books. Consider the object of interest to be of vertical extent h (from A to B) with its base a distance Y above the observer's eye line (the x-axis, with the observer at the origin O). From Figure 3.1 it is clear that the angle $A\hat{O}B = \theta = \alpha - \beta$ is to be maximized. If A has coordinates (x, Y) in general, it follows that

$$\theta = \arctan\frac{Y+h}{x} - \arctan\frac{Y}{x}, \tag{3.1}$$

$$\frac{d\theta}{dx\alpha} = \frac{Y}{x^2 + Y^2} - \frac{Y+h}{x^2 + (Y+h)^2}. \tag{3.2}$$

Therefore an extremum occurs when

$$Y\left[x^2 + (Y+h)^2\right] = (Y+h)(x^2 + Y^2), \text{ i.e., when } x = \left(Y^2 + Yh\right)^{1/2} \equiv X.$$

We know this corresponds to a maximum angle, so we leave it to the reader to verify that $\theta''(X) < 0$. For the Statue of Liberty, the plinth height Y is 47 meters, and the Lady herself is almost as tall: $h = 46$ meters. Therefore if the ground is flat we need to stand at a distance of $X = \left(46 \times 47 + 47^2\right)^{1/2} = 66$ meters.

It is interesting to verify that the points $O, A,$ and B lie on a circle: if this is the case, the center of the circle has coordinates $(0, Y + h/2)$ by symmetry. Then if the radius of the circle is r it follows that

$$x^2 + \left(y - \left[Y + \frac{h}{2}\right]\right)^2 = r^2 \equiv X^2 + \left(\frac{h}{2}\right)^2. \tag{3.3}$$

It is readily verified that the points $A(X, Y), B(X, Y + h)$, and $O(0, 0)$ all satisfy this equation, so the eye level of the observer is tangent to the circle. And another interesting fact is that this maximum angle is attained from any other point on the circle (by a theorem in geometry). Of course, this is only useful to the observer if she can levitate!

This result for the "vertical" circle (Figure 3.1) has implications for a "horizontal" circle in connection with the game of rugby—the best position to place the ball for conversion after a try is on the tangent point of the circle (see the book by Eastaway and Wyndham [8] for more details; they also mention the optimal viewing of two other statues—Christ the Redeemer in Rio de Janeiro and Nelson's Column in London).

$X = T$: THE CONCERT HALL

We have tickets to the symphony. It is well known that the "acoustics" in some concert halls are better than in others—a matter of design. A fundamental question in this regard is—how long does it take for a musical (or other) sound to die out? But this is a rather imprecise question. As a rough rule of thumb, for the "average" (and hence nonexistent) person, weak sound just audible in a quiet room is about a millionth as intense as normal speech or music, so we'll define the reverberation time T in seconds as the time required for a sound to reach 10^{-6} of its original intensity (Vergara 1959). Clearly, the value of T depends on many things, particularly the dimensions of the room and how well sound is absorbed by the wall, floor, ceiling, and the number of people present. Suppose that as the sound is scattered (here meaning reflected and absorbed) and the average distance between reflecting surfaces is L feet. On encountering such a surface suppose that (on average) a proportion a of the intensity is lost. We will call a the *absorptivity* coefficient. If the original sound intensity is I_0, then the intensity after n reflections is

$$I_n \approx (1 - a)^n I_0. \tag{3.4}$$

We'll assume this to be an exact expression from now on. Then we naturally ask: when does $I_n = 10^{-6} I_0$? In other words, if $(1 - a)^n = 10^{-6}$, what is n (to the nearest integer)? It is convenient to work with common logarithms here, so

$$n = \frac{-6}{\log_{10}(1-a)} \tag{3.5}$$

(to the nearest integer). If c (ft/s) is the speed of sound in air at room temperature and pressure, then on average over a distance L, sound will be reflected approximately c/L times per second. This enables us to write the reverberation time T as

$$T = \frac{\text{number of reflections}}{\text{number of reflections per second}} = \frac{n}{c/L} = \frac{nL}{c} = \frac{-6L}{c\log_{10}(1-a)}. \tag{3.6}$$

T is a linear function of L as we would expect, but before putting in some "typical" numbers, let's see how the quantity $N = n/6 = Tc/6L = -\left[\log_{10}(1-a)\right]^{-1}$ behaves as a function of the absorptivity a.

It can be seen from Figure 3.2 and equation (3.6) that both n and T fall off rapidly for absorptivities up to about 20%. In fact, if we use $c \approx 1120$ fps and $L \approx 20$ ft, then $n \approx 270$ and therefore the number of reflections per second is $c/L \approx 56$ s^{-1}. This means that $T \approx 5$ s, which is an awfully long time—almost long enough for the audience to leave the symphony hall in disgust! We need to do better than this. This is why soundproof rooms have walls that look like they are made of egg cartons: the value of a needs to be much higher than 0.05.

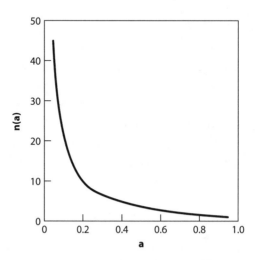

Figure 3.2. The reflection function $n(a)$.

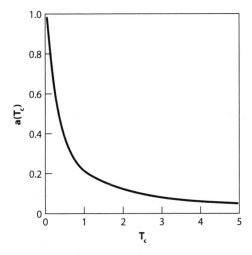

Figure 3.3. Absorptivity as a function of threshold reverberation time T_c.

However, we can invert the problem: suppose that the musicologists and sound engineers have determined that T should be no larger than T_c seconds. With the same values of c, and L, we now require that $n = 56T_c$, so that

$$(1-a)^{56T_c} = 10^{-6}, \text{ or } a = 1 - 10^{-(6/56T_c)}. \tag{3.7}$$

For $T_c = 1$ s, $a \approx 0.22$; for $T_c = 0.5$ s, $a \approx 0.39$, and $T_c = 0.1$ s, $a \approx 0.92$. That's more like it! A graph illustrating the dependence of a on T_c is shown in Figure 3.3. Again, note the rapid fall-off when $T_c < 1$ s.

In reality, the physics of concert halls must be very much more complicated than this, but you didn't really expect that level of sophistication in this book, did you?

X = R: SKYSCRAPERS

Figures 3.4 and 3.5 show, respectively, rebuilding at Ground Zero in New York City, and an office tower at Delft University of Technology in The Netherlands. In both cases we see tall buildings, very much taller than most kinds of trees. But trees sway in the wind, right? So why shouldn't the same be true

Figure 3.4. Rebuilding at the site of Ground Zero, New York City. Photo by Skip Moen.

for tall buildings? In fact, it *is* true. The amplitude of "sway" near the top of the John Hancock building in Chicago can be about two feet. At the time of writing, the world's tallest building has opened in Dubai. The Burj Khalifa stands 828 meters (2,716 ft), with more than 160 stories. It may not be long before even this building is eclipsed by yet taller ones. How much might such a building sway in the wind? Structural engineers have a rough and ready rule: divide the height by 500; this means the "sway amplitude" for the Burj Khalifa is about 5.5 feet! The reason for such swaying is intimately associated with the wind, of course; wind flows around buildings and bridges in a similar fashion to the way water flows around obstacles in a stream. Careful observation of this dynamic process reveals that small vortices or eddies swirl near the obstacle, be it a rock, twig, or half-submerged calculus book. The atmosphere is a fluid, like the stream (though unlike water, it is compressible) and buildings are the obstacles. If wind vortices break off the building in an organized, rhythmic

Figure 3.5. Tower at the Delft University of Technology, Delft, The Netherlands. Photo by the author.

fashion, the building will sway back and forth. Or, to put it another way, a skyscraper (or a smokestack) can behave like a giant tuning fork!

The oscillations arise from the alternate shedding of vortices from opposite sides of the tall structure. Their frequency depends on the wind speed and the size of the building. Not surprisingly, such "flow-induced" oscillations can be very dangerous, and engineers seek to design structures to minimize them [9]. Figure 3.6 illustrates in schematic form how such vortices can develop around a cylindrical body and alternatively "peel off" downstream.

There are two dimensionless numbers that are very important in a study of flow patterns around obstacles. One is called the *Reynolds number*, denoted by R, and is defined by

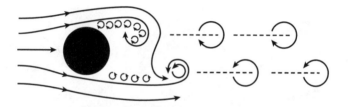

Figure 3.6. Schematic view of vortex formation around a cylindrical obstacle.

$$R = \frac{LU}{\nu},$$

where L and U are characteristic size and (here) wind speed, respectively, and ν is a constant called the kinematic viscosity. For small values of $R(R < 1)$ there is no separation: the cylinder just causes a symmetrical "bump" in flow. For $R > 1$ things get more complicated; in the range $1 < R < 30$ small vortices develop behind the obstacle, but symmetrically about the axis of symmetry, so there is nothing yet to drive oscillations. For $R > 40$ this symmetric flow becomes unstable, and vortices are shed alternately from side to side (as viewed from above). This pattern can exist for Reynolds number up to several thousand (even up to 10^5 if the obstacle is very "smooth").

The other important number is the *Strouhal* number, S. It is named after *Vincenc Strouhal*, a Czech physicist who in 1878 investigated aeolian tones—the "singing" of wires set into oscillation by the wind. Again, in terms of the wind speed U and the diameter d of the wire, S is defined by

$$S = \frac{f_s d}{U},$$

f_s being the frequency of the vortex shedding. Naturally it is to be expected that the Strouhal number is related to the Reynolds number, but for a wide range of the latter, S is almost constant, varying at most between 0.15 and 0.2. When f_s is close to the natural frequency of vibration of the obstacle, the latter can "capture" the former, and the resulting resonance can be very dangerous, for obvious reasons. This phenomenon is called "lock-in."

And speaking of swaying buildings, a few minutes before I wrote this, my office building started to sway. I'm on the second floor, and didn't feel anything, but I heard the bookcases behind me move back and forth. They have a lot of books in them, and are very heavy. Subsequently I heard tales from across campus of swinging lights and moving floors—yes, Virginia, we just had an earthquake! Initial reports indicated it was 5.8 on the Richter scale, with an epicenter midway between Charlottesville and and Richmond, Virginia (and about 80 miles from Washington, D.C.). Tremors were felt as far as New York, Massachusetts, Ohio, Tennessee. and the Carolinas, and the U.S. Capitol and the Pentagon were evacuated. And as I write, Hurricane Irene, currently a Category 3 storm, is making her way steadily toward the Eastern seaboard! Thankfully, by the time it made landfall the storm had weakened to Category 1, but it still caused one death and extensive property damage in the Outer Banks, North Carolina, New Jersey, New York, and Vermont.

$X = d(x,y)$: THE MALL

Note that in the UK, mall = shopping center = shopping centre!

More and more frequently, malls are being located out of towns and cities, perhaps as a single "megastore" or as a mall-like complex; however, many are still built in cities with pedestrian walkways. What follows is a very simplistic model for the competitiveness of two such malls, carried out by considering how one "stacks up" against the other, as measured by the number of trips (N) made per unit interval to a given location.

An important (but rather subjective) question for urban planners is "How attractive is the mall likely to be?" Factors such as the variety of stores within it, ease of access, and parking facilities all contribute to the answer. Essentially, the attractiveness of the mall determines whether one will prefer to travel farther to get there, as opposed to shopping at a nearer but less attractive one. Another question, fundamental for developing a mathematical model for the competition between malls (or specialty stores and shops, for that matter) is "How does N depend on the distance d from the mall?" [10]. Many factors, including those mentioned above, must be contributory, and it is therefore unlikely that N will be a simple function of d. Nevertheless, that is beyond the scope of this book, and we shall content ourselves with a simpler approach to illustrate some basic principles involved.

Obviously we expect N to decrease with distance d, but how rapidly? Another question concerns what we mean by distance here, that is, should we use the standard Euclidean "metric"in the plane, or a modified version, weighted in some manner to account for geographical or social factors? In all likelihood the latter, but again for simplicity we will stick with circular symmetry, consistent with models that appear later in the book. (However, see Appendix 3 for a brief introduction to the so-called *taxicab* or *Manhattan* metric.) To that end, we define a constant "attraction factor" $a_i > 0$ for each mall (here $i = 1, 2$) and write

$$N_i(d) = \frac{a_i}{d^p}, \ p > 0 \tag{3.8}$$

where the distance $d(x, y)$ from a particular mall is calculated using the standard Pythagorean distance formula. The two malls A and B will be located at the coordinate origin and at $(b, 0)$, respectively. We wish to determine the locus of points in the (x, y)-plane such that $N_1 = N_2$. This will be the (closed) boundary curve inside and outside of which one mall is preferred over the other. Therefore we can write

$$\frac{a_1}{(x^2 + y^2)^{p/2}} = \frac{a_2}{((b - x)^2 + y^2)^{p/2}}. \tag{3.9}$$

Rearranging this equation by raising both sides to the power $2/p$ we obtain

$$k\left((b - x)^2 + y^2\right) = \left(x^2 + y^2\right) \tag{3.10}$$

where the positive parameter $k = (a_1/a_2)^{2/p}$. In so doing we have effectively transferred the dependence on the distance (via the parameter p) to a measure of the "relative attractiveness" of the malls. For given values of a_i, note that k increases as p decreases toward zero if $a_1/a_2 > 1$, and k decreases as p decreases toward zero if $a_1/a_2 < 1$. Upon completing the square it follows that for $k \neq 1$

$$\left(x - \frac{kb}{k - 1}\right)^2 + y^2 = \frac{kb^2}{(k - 1)^2}, \tag{3.11}$$

that is, the "boundary of attraction" is a circle of radius $\frac{b\sqrt{k}}{|k - 1|}$ centered at the point $\left(\frac{kb}{k - 1}, 0\right)$. If $0 < k < 1$, that is, mall A is considered less attractive than mall B, the center of the boundary is on the negative x-axis. Furthermore, in this case $\sqrt{k} > k$ so that the circle extends into the half-plane and the point closest to mall B is $\left(\frac{\sqrt{k}b}{1 + \sqrt{k}}, 0\right)$ at a distance $\frac{b}{1 + \sqrt{k}}$ from it. If $k > 1$, mall A is the more attractive of the two, and the closest point on the boundary to the point A is also a distance $\frac{b}{1 + \sqrt{k}}$ from it. Figure 3.7 illustrates this case for arbitrary $k > 1$; the dotted

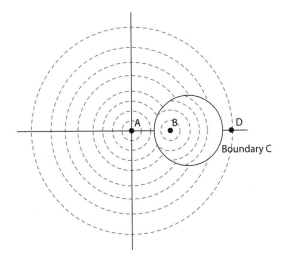

Figure 3.7. Equidistance contours and "attraction" boundary for a circular city.

contours represent circular arcs on which (i) N_1 is constant (outside the solid circular boundary) and (ii) N_2 is constant (inside the boundary), respectively.

One obvious and simple change to the model would be to consider elliptical contours of constant N_i; thus for the equation of the boundary curve we might use

$$\frac{a_1}{(x^2 + a^2 y^2)^{p/2}} = \frac{a_2}{((b-x)^2 + a^2 y^2)^{p/2}} \tag{3.12}$$

instead of equation (3.9). In this case, if $a > 1$ the loci of constant N_i will be ellipses with semi-major axes along the x-axis, and the resulting boundary curve is also an ellipse, not surprisingly, with the equation

$$\left(x - \frac{kb}{k-1}\right)^2 + a^2 y^2 = \frac{kb^2}{(k-1)^2}. \tag{3.13}$$

A representative diagram is shown in Figure 3.8.

X = P: THE POST OFFICE

I know that it's usually a bad idea to go to the post office to mail letters when it first opens, especially on a Monday morning. And around the middle of December

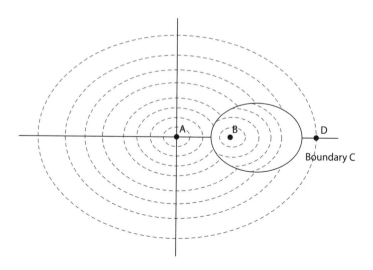

Figure 3.8. Equidistance contours and "attraction" boundary for an elliptical city.

it's a nightmare! But sometimes there is little or no line at all, even when the office is normally busy. That has happened on my last two visits. And conversely, sometimes there is a long line when I least expect it. While the average number of customers may be one every couple of minutes, it is most unlikely that each one will arrive every two minutes: there will be a certain "clumpiness" in the arrivals. The *Poisson distribution* describes this clumpiness well. Details and applications of this distribution are discussed in Chapter 9 and in Appendix 4, but we will summarize the results here. If the arrivals at the post office, checkout line, traffic line, busstops, and so on are random and average to λ per minute (or other unit of time), then the probability $P(n)$ of n customers arriving in any given minute is

$$P(n) = \frac{\lambda^n e^{-\lambda}}{n!}. \tag{3.14}$$

Suppose that on a reasonably busy day at the post office, $\lambda = 2$. What is the chance that there will be six customers ahead of me when I arrive? (Of course, there is also the question of how quickly on average the counter clerks serve the customers, but that is another issue.)

Then

$$P(6) = \frac{2^6 e^{-2}}{6!} \approx 0.012,$$

or just over 1 percent. That's not bad! And that's just for starters; we will meet the Poisson distribution again in Chapter 9.

$X = Pr$: CHANCE ENCOUNTER?

Another probability problem. Suppose that Jack and Jill, freshly arrived in the Big City, each have their own agendas about things to do. They decide—sort of—to do their "own thing" for most of the day and then meet sometime between 4:00 p.m. and 6:00 p.m. to have a bite to eat and discuss the day's activities. Both of them are somewhat disorganized, and they fail to be more specific than that (after all, there are *so* many books to look at in Barnes and Noble, or Foyle's Bookshop in London, who knows when Jack will be ready to leave?) Furthermore, neither cell (mobile) phone is charged, so they cannot ask "where the heck are you?"

Suppose that each of them shows up at a random time in the two-hour period. Jack is only willing to wait for 15 minutes, but Jill, being the more patient of the two (and frankly, a nicer person) is prepared to wait for half an hour. What is the probability they will actually meet? The related question of what will happen if they don't is beyond the scope of this book, but the first place to start looking might be the local hospital.

This is an example of geometric probability. In the 2-hr square shown in Figure 3.9, the diagonal corresponds to Jack and Jill arriving at the same time at any point in the 2-hr interval. The shaded area around the diagonal represents the "tolerance" around that time, so the probability that they meet is the ratio of the shaded area to the area of the square, namely,

$$Pr = \frac{1}{4}\left[4 - \frac{1}{2}\left(2 - \frac{1}{4}\right)^2 - \frac{1}{2}\left(2 - \frac{1}{2}\right)^2\right] \approx 0.336,$$

or about one third. Not bad. Let's hope it works out for them.

$X = Q(L, M)$: BUILDING IN THE CITY

To build housing requires land (L) and materials (M), and if the latter consists of everything that is not land (bricks, wood, wiring, etc.), then we can define a function $Q(L,M)$,

$$Q(L,M) = aL^\lambda M^{1-\lambda}, \tag{3.15}$$

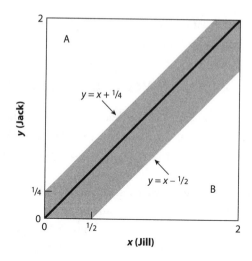

Figure 3.9. The shaded area relative to that of the square is the probability that Jack and Jill meet.

where $a > 0$ and $0 \le \lambda \le 1$. This form of function is known as a *Cobb-Douglas* production function, widely used in economics, and named for the mathematician and U.S. senator who developed the concept in the late 1920s.

If the price of a unit of housing is p, the cost of a unit of land per unit area is l, and the cost of unit materials is m (however all these units may be defined), then the builders' total profit P is given by

$$P = pQ - lL - mM. \tag{3.16}$$

If a speculative builder wishes to "manipulate" L and M to maximize profit, then at a stationary point

$$p\frac{\partial Q}{\partial L} = l \text{ and } p\frac{\partial Q}{\partial M} = m.$$

Exercise: (i) Show, using equation (3.16), that this set of equations takes the algebraic form

$$p\lambda Q = lL; \; p(1-\lambda)Q = mM. \tag{3.17}$$

(ii) Show also from equation (3.16) that according to this model, such a speculator "gets his just deserts" (i.e., $P = 0$)!

We shall use equations (3.15) and (3.17) to eliminate Q, L, and M to obtain

$$p = C(\alpha,\lambda)l^{\lambda}m^{1-\lambda}, \tag{3.18}$$

where $C(\alpha,\lambda)$ is a constant. With two additional assumptions this will enable us to derive a valuable result to be used in Chapter 17 (The axiomatic city). If the cost of materials is independent of location, and land costs increase with population density ρ according to a power law, that is, $l \propto \rho^{\kappa}(\kappa > 0)$, then

$$p = A\rho^{\gamma},$$

where $\gamma = \kappa\lambda$ and A is another constant.

Exercise: Derive equation (3.18)

We conclude this section with a numerical application of equation (3.15) in a related context. It is recommended that the reader consult Appendix 5 to brush up on the method of *Lagrange multipliers* if necessary.
Suppose that

$$Q(L,M) = 16L^{1/4}M^{3/4}, \tag{3.19}$$

L now being the number of labor "units" and M the number of capital "units" to produce Q units of a product (such as sections of custom-made fencing for a new housing development). If labor costs per unit are $l = \$50$, capital costs per unit are $m = \$100$, and a total of $\$500,000$ has been budgeted, how should this be allocated between labor and capital in order to maximize production, and what is the maximum number of fence sections that can be produced? (I hope the reader appreciates the "nice" numbers chosen here.)
To solve this problem, note that the total cost is given by

$$50L + 100M = 500,000,$$
$$\text{i.e., } L + 2M = 10,000. \tag{3.20}$$

The simplified equation will be the constraint used below. The problem to be solved is therefore to maximize Q in equation (3.19) subject to the constraint (3.20). Using the method of Lagrange multipliers we form the function

$$F(L,M;\lambda) = 16L^{1/4}M^{3/4} + \lambda(L+2M-10,000). \tag{3.21}$$

The critical points are found by setting the respective partial derivatives to zero, thus:

$$\frac{\partial F}{\partial L} = 4L^{-3/4}M^{3/4} + \lambda = 0; \tag{3.22a}$$

$$\frac{\partial F}{\partial M} = 12L^{1/4}M^{-1/4} + 2\lambda = 0; \tag{3.22b}$$

$$\frac{\partial F}{\partial \lambda} = L + 2M - 10{,}000 = 0. \tag{3.22c}$$

From equations (3.22a) and (3.22b) we eliminate λ to find that $M = 3L/2$. From (3.22c) it follows that $M = 2500$, and hence that $L = 3750$. Finally, from (3.22a), $\lambda = -4(2500)^{-3/4}(3750)^{3/4} \approx -5.4216$, so the unique critical point of F is $(2500, 3750, -5.4216)$. The maximum value is $Q(L, M) = 16(2500)^{1/4}$ $(3750)^{3/4} \approx 54{,}200$ sections of fencing.

Question: Can we be certain that this is the *maximum* value of Q?

Exercise: Show that had we not simplified the constraint equation (3.20), the value for λ would have been $\lambda \approx -0.1084$ but the maximum value of Q would remain the same.

$X = A_n$: A MORTGAGE IN THE CITY

The home has been built; now it's time to start paying for it. It's been said that if you think no one cares whether you're alive or not, try missing a couple of house payments. Anyway, let's set up the relevant *difference equation* and its solution. It is called a difference equation (as opposed to a *differential* equation) because it describes discrete payments (as opposed to continuous ones).

Suppose that you have borrowed (or currently owe) an amount of money A_0, and the annual interest (assumed constant) is $100I\%$ per year (e.g., if $I = 0.06$, the annual interest is 6%), compounded m times per year. If you pay off an amount b each compounding period (or due date), the governing first-order nonhomogeneous difference equation is readily seen to be

$$A_{n+1} = \left(1 + \frac{I}{m}\right)A_n - b. \tag{3.23}$$

The solution is

$$A_n = \left(A_0 - \frac{bm}{I}\right)\left(1 + \frac{I}{m}\right)^n + \frac{bm}{I}. \tag{3.24}$$

Let's see how to construct this solution using the idea of a fixed point (or equilibrium value). Simply put, a fixed point in this context is one for which $A_{n+1} = A_n$. From equation (3.23) such a fixed point (call it L) certainly exists: $L = Bm/I$. If we now define $a_n = A_n - L$, then it follows from equation (3.24) that

$$a_{n+1} + L = \left(1 + \frac{I}{m}\right)(a_n + L) - L\frac{I}{m} = \left(1 + \frac{I}{m}\right)a_n + L,$$

that is

$$a_{n+1} = \left(1 + \frac{I}{m}\right)a_n \equiv \lambda a_n. \tag{3.25}$$

Thus we have reduced the nonhomogeneous difference equation (3.23) to a homogeneous one. Since $a_1 = \lambda a_0, a_2 = \lambda a_1 = \lambda^2 a_0, a_3 = \lambda a_2 = \lambda^3 a_0$, etc., it is clear that equation (3.25) has a solution of the form $a_n = \lambda^n a_0$. On reverting to the original variable A_n and substituting for L and λ the solution (3.24) is recovered immediately.

Exercise: Verify the solution (3.24) by direct substitution into (3.23).

Here is a natural question to ask: *How long will it be before we pay off the mortgage?* The answer to this is found by determining the number of pay periods n that are left (to the nearest preceding integer; there will generally be a remainder to pay off directly). The time (in years) to pay off the loan is just n/m (usually $m = 12$ of course). To find it, we set $A_n = 0$ and solve for n. Thus

$$n = -\frac{\log\left(1 - \frac{A_0 I}{bm}\right)}{\log\left(1 + \frac{I}{m}\right)}. \tag{3.26}$$

Exercise: Verify this result. And use it to find out when you will have paid off *your* mortgage!

Chapter 4

EATING IN THE CITY

The first time I visited a really big city on more than a day trip, I wondered where people found groceries. I had just left home to start my time as an undergraduate in London, but at home I could either ride my bicycle or use my father's car to drive to the village store or visit my friends (not always in that order). In London I used the "tube"—the underground transport system (the subway) and sometimes the bus, but I walked pretty much everywhere else. I soon found out where to get groceries, and in retrospect realized that there are some quite interesting and amusing mathematics problems associated with food, whether it's in a city or in the middle of the country. What follows is a short collection of such items connected by a common theme—eating!

X = W: WATERMELON WEIGHT

A farmer harvested ten tons of watermelons and had them delivered by truck to a town 30 miles away. The trip was a hot and dusty one, and by the time the destination was reached, the watermelons had dried out somewhat; in fact their water content had decreased by one percentage point from 99% water by weight to 98%.

Question: What was the weight of the watermelons by the time they arrived in the town?

$W_1 = 0.99W_1 + 0.01W_1$ (water weight plus pith weight) and

$W_2 = 0.98W_2 + 0.02W_2$, but the pith weight is unchanged; therefore $0.01W_1 = 0.02W_2$, and so

$W_2 = W_1/2$;

the weight of the watermelons upon arrival is now only five tons. Very surprising, but it shows that a small percentage of a large number can make quite a difference . . .

X = V: HOW MUCH OF THAT FRUIT IS FRUIT?

Suppose that Kate feels like having a healthy snack, and decides to eat a banana. Mathematically, imagine it to be a cylinder in which the length L is large compared with the radius r. Suppose also that the peel is about 10% of the radius of the original banana. Since the volume of her (now) idealized right circular banalinder (or cylinana) is $\pi r^2 L$, she loses 19% of the original volume when she peels it $(1 - (0.9)^2 = 0.81)$. Okay, now let's do the same thing for a spherical orange of volume $4\pi r^3/3$. The same arguments with the peel being about 10% of the radius give a 27% reduction in the volume $(1 - (0.9)^3 = 0.73)$. We might draw the conclusion in view of this that it is not very cost-effective to buy bananas and oranges, so she turns her stomach's attention to a peach. Now we're going to ignore the thickness of the peach skin (which I eat anyway) in favor of the pit. We'll assume that it is a sphere, with radius 10% of the peach radius. Then the volume of the pit is 10^{-3} of the volume of the original peach; a loss of only 0.1%. What if it were 20% of the peach radius? The corresponding volume loss would be only 0.8 %. These figures are perhaps initially surprising until we carry out these simple calculations [11]. But is a banana a fruit or an herb? Inquiring minds want to know.

Meanwhile, the neighbor's hotdogs are cooking. How much of the overall volume of a hotdog is the meat? Consider a cylindrical wiener of length L and radius r surrounded by a bun of the same length and radius $R = ar$, where $a > 1$. If the bun fits tightly, then its volume is

$$V_b = \pi L(R^2 - r^2) = \pi L r^2(a^2 - 1) = (a^2 - 1)V_m, \tag{4.1}$$

where V_m is the volume of the wiener. If $a = 3$, for example, then $V_b = 8V_m$. But a cheap hotdog bun is mostly air; about 90% air in fact!

$X = t_c$: TURKEY TIME

Question: How long does it take to cook a turkey (without solving an equation)?

Let's consider a one-dimensional turkey; these are difficult to find in the grocery store. Furthermore, you may object that a spherical turkey is much more realistic than a "slab" of turkey, and you'd be correct! A spherical turkey might be a considerable improvement. However, the equation describing the diffusion of heat from the exterior of a sphere (the oven) to the interior can be easily converted by a suitable change of variables to the equation of a slab heated at both ends, so we'll stick with the simpler version.

The governing equation is the so-called heat or diffusion equation (discussed in more detail in Appendix 10)

$$\frac{\partial T}{\partial t} = \kappa \frac{\partial^2 T}{\partial x^2}, \ 0 < x < L, \tag{4.2}$$

where T is the temperature at any distance x within the slab at any time t; κ is the coefficient of thermal diffusivity (assumed constant), which depends on the thermal properties and density of the bird; and L is the size (length) of the turkey. This equation, supplemented by information on the temperature of the turkey when it is put in the oven and the oven temperature, can be solved using standard mathematical tools, but the interesting thing for our purposes is that we can get all the information we need without doing that. In this case, the information is obtained by making the equation above dimensionless. This means that we define new variables for which (i) the dimensions of time and length are "canceled," so to speak, and (ii) the temperature is defined relative to the interior temperature of the bird when it is fully cooked. We'll call this temperature T_c. We'll also define t_c as the time required to attain this temperature T_c—the cooking time. It is this quantity we wish to determine as a function of the size of the bird. The advantage of this formulation is that we don't have to repeat this calculation for each and every turkey we cook: indeed, as we will see, with a little more sophistication we can express the result in terms that are independent of the size of the turkey.

To proceed with the "nondimensionalization" let $T' = T/T_c$, $t' = t/t_c$, and $x' = x/L$. Using the chain rule for the partial derivatives, equation (4.2) in the new variables takes the form

$$\frac{\partial T'}{\partial t'} = \frac{\kappa t_c}{L^2}\frac{\partial^2 T'}{\partial x'^2}, \quad 0 < x' < 1. \tag{4.3}$$

Has anything at all been accomplished? Indeed it has. Since both derivatives are expressed in dimensionless form, then so must be the constant $\kappa t_c / L^2$. Let's call this constant a. It follows that

$$t_c = \frac{aL^2}{\kappa}, \text{ i.e. } t_c \propto L^2.$$

But the weight W of anything is (for a given mean density) proportional to its volume, and its volume is proportional to the cube of its size, that is, $W \propto L^3$, so $L^2 \propto W^{2/3}$, from which we infer that $t_c \propto W^{2/3}$. And that is our basic result: *the time necessary to adequately cook our turkey is proportional to the two-thirds power of its weight.* We have in fact made use of a very powerful technique in applied mathematics in general (and mathematical modeling in particular): *dimensional analysis.* As seen above, this involves scaling quantities by characteristic units of a system, and in so doing to reveal some fundamental properties of that system.

Where can we go from here? One possible option is to determine the unknown constant of proportionality a/κ; in principle κ can be found (but probably not in any cookbook you possess), but of course a is defined in terms of t_c, which doesn't help us! However, if we have a "standard turkey" of weight

W_0 and known cooking time t_0, then for any other turkey of weight W, a simple proportion gives us

$$t_c = \left(\frac{W}{W_0}\right)^{2/3} t_0. \qquad (4.4)$$

From this the cooking time can be calculated directly. Note that t_c is not a linear function of weight; in fact the cooking time *per pound* of turkey decreases as the inverse cube root of weight, since

$$\frac{t_c}{W} = \frac{t_0}{W^{1/3} W_0^{2/3}}.$$

Hence doubling the cooking time t_0 for a turkey of weight $2W_0$ may result in an overcooked bird; the result (4.4) implies that $t_c = 2^{2/3} t_0 \approx 1.6 t_0$ should suffice. However, always check the bird to be on the safe side. Note that the units used here, pounds, are really pounds-force, a unit of weight. Generally, pounds proper are units of mass, not weight, but in common usage the word is used to mean weight.

X = N: HOW MANY TOMATOES ARE CONSUMED BY CITY-DWELLERS EACH YEAR?

Of course, this is a rather ill-posed question. Are we referring to large tomatoes, which we slice nd put in salads, or those little ones that we find in salad bars? Of course, we can find both sizes, and those in between, in any supermarket or produce store. And I suspect that more tomatoes are consumed during the summer than the rest of the year, for obvious reasons. (Note that we are not including canned tomatoes used in pasta sauce.) Now big tomatoes and little tomatoes share a very import characteristic: they are both tomatoes! I'm going to work with a typical timescale of one week; that is, I might use one large tomato per week in my sandwiches, or consume more of the smaller tomatoes in the same period of time. Therefore I will take the average in the following sense. A fairly big tomato 3 inches in diameter is about 30 times as large by volume as one that is one inch in diameter, so we can use the Goldilocks principle referred to earlier—*is it too large, too small, or just right?* To implement this, we merely take the geometric mean of the volumes. (A reminder: the geometric mean of two positive numbers is the most appropriate measure of "average" to take when the numbers differ by orders of magnitude.) The geometric mean of 1 and 30 is about 5, so accounting for the range of sizes, we'll work with 5 "generic" tomatoes eaten per week in the calculations below. Let's suppose that about a third of the U.S. population of 300 million eat tomatoes regularly, at least during the summer. (I am therefore ignoring those adults and children who do not eat them by dividing the population into three roughly equal groups, again using the Goldilocks principle: fewer than 100% but more than 10% of the population eat tomatoes; the geometric mean is $\sqrt{100 \times 10} \approx 30\,(\%)$, or about 1/3.)

Therefore, if on average one third of the population eat 5 tomatoes per week, then halving this to reflect a smaller consumption (probably) during the winter months (except in Florida ☺), the approximate number of tomatoes eaten every year is

$$N \approx \frac{1}{3} \times 3 \times 10^8 \times \frac{1}{2} \times 50 \approx 3 \times 10^9,$$

that is, three billion tomatoes. We multiply this by about 2/3 (to account for the proportion of city-dwelling population in the U.S.) to get roughly two billion (with no offense to non-city dwellers, I trust).

X = Pr: PROBABILITY OF BITING INTO . . .

I love apples, don't you? Sometimes though, they contain "visitors." Suppose that there is a "bug" of some kind in a large spherical apple of radius, say, two inches. We will assume that it is equally likely to go anywhere within the apple (we shall ignore the core). What is the probability that it will be found within a typical "bite-depth" of the surface? Based on my lunchtime observations, I shall take this as ¾ inch, but as always, feel free to make your own assumptions.

The probability P of finding the bug within one bite-depth of the surface is therefore the following ratio of volumes (recall that the volume of a sphere is $4\pi/3$ times the cube of its radius):

$$P = \frac{\text{volume of the spherical shell}}{\text{volume of the apple}} = \frac{4\pi/3}{4\pi/3} \times \frac{\left[2^3 - \left(2 - \frac{3}{4}\right)^3\right]}{2^3} = \frac{8 \times \frac{125}{64}}{8}$$

$$= \frac{387}{512} \approx 0.76,$$

or about 76%. *Ewwwhhh!*

Chapter 5

GARDENING IN THE CITY

When I was growing up, I loved to visit my grandfather. Despite living in a city (or at least, a very large town) he was able to cultivate and maintain quite a large garden, containing many beautiful plants and flowers. In fact, whenever I asked what any particular flower was called, his reply was always the same: "*Ericaceliapopolifolium!*" I never did find out whether he was merely humoring me, or whether he didn't know! In addition to his garden, he rented a smaller strip of land (an "allotment") farther up the road where, along with several others, he grew potatoes, carrots, beans, and other vegetables. I'm ashamed to admit that I didn't inherit his love for the art of gardening, much to my parents' disappointment. It skipped a generation, though; my son has a lovely garden and my son-in-law has a very "green thumb" (of course, neither I nor my grandfather can be held responsible for the latter).

As with the previous chapter, there are various and sundry topics in this one, connected (somewhat tenuously, to be sure) by virtue of being found in a garden or greenhouse. Let's get specific. Plants grow. My daughter and her family live in northern Virginia, and at the bottom of their garden they have quite a lot of bamboo plants. These can reach great heights, so my question is,

$X = h'(t)$: **Question**: How fast does bamboo grow?

The growth rate for some types of bamboo plant may be as much as 4 meters (about 13 ft) per day. Let's work with a more sedate type of bamboo, growing only (!) at the rate of 3 ft/day (about 1 m / day). Just for fun, let's convert this rate to (i) miles/second, (ii) mph, and (iii) km/decade (ignoring leap years!).

$$\text{(i)} \quad 3\frac{\text{ft}}{\text{day}} = 3\frac{\text{ft}}{\text{day}} \times \frac{1 \text{ mile}}{5280 \text{ ft}} \times \frac{1 \text{ day}}{24 \text{ hr}} \times \frac{1 \text{ hr}}{60 \text{ min}} \times \frac{1 \text{ min}}{60 \text{ s}} \approx 7 \times 10^{-9} \text{ mps.}$$

$$\text{(ii)} \quad 3\frac{\text{ft}}{\text{day}} = 3\frac{\text{ft}}{\text{day}} \times \frac{1 \text{ mile}}{5280 \text{ ft}} \times \frac{1 \text{ day}}{24 \text{ hr}} \approx 2 \times 10^{-5} \text{ mph.}$$

$$\text{(iii)} \quad 3\frac{\text{ft}}{\text{day}} = 3\frac{\text{ft}}{\text{day}} \times \frac{12 \text{ inches}}{1 \text{ ft}} \times \frac{2.54 \text{ cm}}{1 \text{ inch}} \times \frac{1 \text{ m}}{100 \text{ cm}} \times \frac{1 \text{ km}}{1000 \text{ m}} \times \frac{3650 \text{ days}}{1 \text{ decade}}$$
$$\approx 3 \text{ km/decade.}$$

As I write this, my daughter and son-in-law have sold their house and moved into a somewhat larger one. Could it be that the bamboo drove them out?

Grass also has a propensity to grow, but it seems that my lawn is never as green and lush as everyone else's. Despite my secret desire to cover it with Astroturf, or green-painted concrete, I bought yet another bag of grass seed awhile ago. It can be thought of as a rectangular box with dimensions approximately $2 \times 1.5 \times 0.5 \text{ ft}^3$. So let's think about the quantity of seed in this bag . . .

$X = N$: PROBLEM

Estimate (i) the number of seeds in such a bag (assumed full), and (ii) how many such bags would be required to seed, say, a golf course with area one square mile.

(i) We need first to estimate the volume of a typical grass seed. Since I started this question using the more familiar British units (for those in the U.S. at least), as opposed to the easier metric units, I'll continue in this vein. Examining a seed, I estimate a typical seed to be a rectangular box with approximate dimensions $1/5$ in $\times$ $1/20$ in $\times$ $1/20$ in, or about $1/2000$ cu in. The bag's volume is about $25 \times 20 \times 5 = 2500$ cu in. Therefore the number of seeds is $N = 2500 \div (1/2000) = 5 \times 10^6$, or 5 million! How many seeds would there be to seed one square mile?

(ii) Let's assume that the grass is seeded uniformly with about 10 to 20 seeds per sq in. (you can change this seed density to suit your own estimates). I'll go with the lower figure. Since a (linear) mile contains $5280 \times 12 \approx 6 \times 10^4$ inches, a square mile contains the square of this number, and with 10 seeds/sq in we have the quantity of seeds as approximately $10 \times (6 \times 10^4)^2 \approx 4 \times 10^{10}$ (that's 40 billion!). Finally, dividing this figure by the average number of seeds per bag we obtain $4 \times 10^{10} \div 5 \times 10^6$, or about 10,000 bags. I could round this up because the first figure is unlikely to be correct (e.g., it could be 6 or 9, and still be within the order of accuracy we anticipate). I wonder what the local golf club would say.

At the university where I am employed, there is a lovely collection of orchids, some of them very rare. Suppose that a botanist (let's call him Felix) wishes to grow one of these rarer varieties in the greenhouse. From his previous attempts, he has concluded that the probability that a given bulb will mature is about one third. He decides to plant six bulbs.

$X = Pr$: **Question**: What is the probability that at least three bulbs will mature?

Since the probability of (i) at least three bulbs maturing *and* (ii) the probability of none or one or two maturing must add up to one (meaning certainty that either outcome (i) or outcome (ii) will occur), then the probability we require is given by

$$P(\geq 3) = 1 - P(0) - P(1) - P(2). \tag{5.1}$$

In order to proceed we need to introduce the concept of *combinations*, $_nC_r$, which represents the number of ways of choosing r items from a total of n (without regard to order). This is defined as

$$_nC_r = \frac{n!}{(n-r)!\,r!},$$

where for a positive integer n, for example, 5, $5! = 5 \times 4 \times 3 \times 2 \times 1 = 120$. Furthermore, by definition, $0! = 1$. Now $P(r)$ is the probability that only r events will occur, and is given by the number of ways the event can occur, multiplied by the probability that r events do occur and $(n - r)$ do not. Thus

$$P(0) = {_6}C_6 \times \left(1 - \left[\frac{1}{3}\right]\right)^6 = \frac{6!}{0!6!} \times \left(\frac{2}{3}\right)^6 = 1 \times \frac{64}{729} \approx 0.09,$$

$$P(1) = {_6}C_1 \times \left(\frac{1}{3}\right) \times \left(1 - \left[\frac{1}{3}\right]\right)^5 = \frac{6!}{1!5!} \times \frac{1}{3} \times \left(\frac{2}{3}\right)^5 = \frac{6 \times 32}{729} \approx 0.26, \text{ and}$$

$$P(2) = {_6}C_2 \times \left(\frac{1}{3}\right)^2 \times \left(1 - \left[\frac{1}{3}\right]\right)^4 = \frac{6!}{2!4!} \times \left(\frac{1}{3}\right)^2 \times \left(\frac{2}{3}\right)^4 = \frac{6 \times 5}{2} \times \left(\frac{16}{729}\right)$$

$$= \frac{15 \times 16}{729} \approx 0.33, \text{ so}$$

$$1 - P(0) - P(1) - P(2) \approx 1 - 0.09 - 0.26 - 0.33 = 0.32.$$

Or just under one third. Go for it, Felix!

According to Richard F. Burton [12], an expert in earthworms (and a contemporary of Charles Darwin) estimated that in a typical field there are about 133,000 earthworms per hectare. (A hectare is a square hectometer; a hectometer is 100 meters, so you'll be delighted to know that we're back with the metric system.)

X=*N*: **Question**: What is this figure in worms per square meter?

Clearly, a square hectometer is 10^4 sq m, so the only thing we need to do is divide by this number to get about 13 worms per square yard. Does that sound about right? Of course, it depends very much on the kind of soil, and we don't worry about the 0.3 earthworm (unless it ends up in our apple [see Chapter 4]).

While we're on the subject of slimy things, consider this. A certain species of slug is 80% water by weight. Suppose that it loses a quarter of this water by evaporation.

X=*W*: **Question**: What is its new percentage of water by weight?

If W and W_{new} are the weights of the slug "before and after," so to speak, then in each case the weight can be distributed as the sum of the water content and the rest, that is,

$W = 0.8W + 0.2W$, so that $W_{new} = 0.75 \times (0.8W) + 0.2W = 0.6W + 0.2W = 0.8W$.

Therefore the percentage of water is now $\dfrac{0.6W}{0.8W} = 75\%$.

This is a little like the watermelon problem, isn't it? Don't confuse the two when eating, though. Do you recall the quote (from St. John of Patmos) about leaves on a tree? Let's set up a simple framework for estimating the number of leaves on any tree or bush.

X=N: **Question**: How many leaves are on that laurel bush in my back yard?

I'll approximate my smallish bush by a sphere of radius half a meter, so that's a surface area (4π times radius squared) of about 3 m². Now the leaves does not "continuously" cover the surface, but then again, there are leaves throughout much of the bush, not just on the outer "canopy," so I'll simplify this problem crudely by just assuming a continuous outer surface composed of leaves about 1 cm square, that is, of area 1 cm² $= 10^{-4}$ m². Dividing 3 m² by this quantity gives us about $N \approx 3 \times 10^4$ leaves.

Doubling the diameter of the bush to make it a small tree would quadruple the area, but the area of the individual leaves would probably be larger (depending on the type of tree), so for a yew tree, say, with typical leaf area about 4 cm², these two effects would essentially cancel each other out, giving us a figure again of about 30,000 leaves, accurate to within a factor of two or three, I suspect.

Exercise: Estimate the number of leaves on that really big tree in your neighborhood. And when you've done that, estimate the total length of the tree; that is, the trunk plus all the branches and twigs.

Chapter 6

SUMMER IN THE CITY

Question: How many squirrels live in Central Park?

Central Park in New York City runs from 59th Street to 110th Street [6]. At 20 blocks per mile, this is 2.5 miles. Central Park is long and narrow, so we will estimate its width at about 0.5 mile. This gives an area of about 1 square mile or about 2 square kilometers.

It's difficult to estimate the number of squirrels in that large an area, so let's break it down and think of the area of an (American) football field (about 50 yards × 100 yards). There will be more than 1 and fewer than 1000 squirrels living there, so we choose the number geometrically between 1 and 1000, or 30 (using the Goldilocks principle again).

This is where the metric system comes in handy. One kilometer (1000 m or about 1100 yards; we'll round down to 1000 yards) is the length of ten football fields and the width of twenty. This means that there are 200 football fields in a square kilometer. Now the number of squirrels in Central Park is about $N = (2 \text{ km}^2) \times (200 \text{ football fields}/(\text{km}^2)) \times (30 \text{ squirrels/football field}) = 12,000 \approx 10^4$. That's enough squirrels for even the most ambitious dog to chase!

$X = X$: SUNBATHING IN THE CITY

It's a lovely weekend and you decide to "catch some rays" in the park. If you're anything like me, with the fair skin of someone from Northern Europe, you will make certain you lather yourself with sun block. It's tempting to get the highest possible SPF variety, but is it really necessary?

In fact, SPF 30 does *not* block out twice as much harmful radiation as SPF 15. SPF is not sun-filtering, it is a S(un)P(rotecting)F(actor). The label tells you how much *time* you can spend in the sun before you start to burn (compared with the time for bare skin); 15 times longer for SPF 15, 30 times longer for SPF 30. However, SPF 30 only blocks out about 3% more of the harmful UVA and UVB radiation.

Here's what happens. Let's suppose we have SPF 2 (does that exist?). Anyway, that would block out 50% of the UV radiation that causes burning. If you burn after 30 minutes when naked as the day you were born (without any sunscreen), you could stay out for an hour—twice as long—with SPF 2. (Please try any naked sunbathing at home, not in the park.) SPF 4 would block out 75% of the harmful radiation, so you could stay out in the sun four times longer—but it cuts out only 25% more of the incoming UV rays than SPF 2 does, correct? SPF 8 means you can stay out 8 times longer, but only cuts out, well, let's see how much more.

SPF 2 cuts out $1 - \frac{1}{2} = 50\%$ of the incoming UV radiation.

SPF 4 cuts out $1 - \frac{1}{4} = 75\%$ of the incoming UV radiation.

SPF 8 cuts out $1 - \frac{1}{8} = 87.5\%$ of the incoming UV radiation.

SPF 16 cuts out $1 - \frac{1}{6} = 93.75\%$ of the incoming UV radiation.

You get the idea. Following this pattern, we see that SPF X cuts out a fraction $1 - \frac{1}{X}$ of the incoming UV radiation. But we can examine this from another point of view. From the list above we see that the fractions of UV radiation blocked by SPF 2, 4, 8, 16 can be written respectively as

$$1 - \frac{1}{2}, 1 - \frac{1}{2^2}, 1 - \frac{1}{2^3}, 1 - \frac{1}{2^4}.$$

Therefore, for SPF X the corresponding fraction of blocked radiation may be written in the form $1 - \frac{1}{2^n}$, where $X = 2^n$. Taking logarithms to base 2 we find that $n = \log_2 X$, or, using the change of base formula with common logarithms, $n = \frac{\log X}{\log 2}$.

Let's now go back and compare $X = 15$ and 30.

For SPF 15, $n = \frac{\log 15}{\log 2} \approx \frac{1.176}{0.301} \approx 3.9$, so $1 - \frac{1}{2^{3.9}} \approx 0.933$ or 93.3% of the harmful radiation is blocked. Of course, we could have just calculated $1 - \frac{1}{15} \approx 0.933$ to get this result—but where would be the fun in that?

For SPF 30, $n = \frac{\log 30}{\log 2} \approx \frac{1.477}{0.301} \approx 4.9$, so $1 - \frac{1}{2^{4.9}} \approx 0.967$ or 96.7% of the harmful radiation is blocked. This is a difference of 3.4 percentage points! Again, "we don't need no logs" to do this, because $1 - \frac{1}{30} \approx 0.967$.

Just for fun, consider SPF 100 (if that exists!).

In this case $n = \frac{\log 100}{\log 2} \approx \frac{2}{0.301} \approx 6.6$, so $1 - \frac{1}{2^{6.6}} \approx 0.99$ or 99% of the harmful radiation is blocked. But you knew that, because $1 - \frac{1}{100} \approx 0.99$.☺

$X = t$: JOGGING IN THE CITY

You are walking in a city park (Central Park, for example; but be careful not to trip over the squirrels). Suppose that a jogger passes you. As she does so, at an average speed, say of 8 mph, you wonder if there is an instant when her speed is *exactly* 8 mph, or equivalently, 7.5 minutes per mile. Fortunately, you have taken a calculus class, and you recall the mean value theorem, an informal (and imprecise) rendering of which says that for a differentiable function $f(t)$ on a closed interval there is at least one value of t for which the tangent to the graph is parallel to the chord joining its endpoints (why not sketch this?). This means that if *f* is the distance covered as a function of time *t*, then at some point (or points) on the run her speed will be exactly 8 mph. A formal statement of the theorem can be found in any calculus book when you get home (unless you are carrying it while walking in the park).

The jogger's name is Lindsay, by the way; by now you've seen each other so often that you greet each other as she races past. As she does so yet again, a related question comes to mind: does Lindsay cover any one continuous mile in exactly 7.5 minutes? (Of course, this assumes her run is longer than a mile.) This is by no means obvious, to me at least, and it may not have occurred to Lindsay either.

Suppose that $t(x)$ is a continuous function representing the time taken to cover *x* miles. We consider this question in two parts. Suppose further that Lindsay runs an integral number (*n*) of miles. If she averages 7.5 min/mi (8 mph), then $t(x) - 7.5x = 0$ when $x = 0$ and $x = n$, i.e. $t(n) - t(0) = 7.5n$. If she never covered any continuous mile in 7.5 minutes, then it follows that the new function $T(x) = t(x+1) - t(x) - 7.5$ is continuous and never zero. Suppose that it is always positive; a similar argument applies if it is always negative. Hence for $x = 0, 1, 2, \ldots, n$, we can write the following sequence of *n* inequalities: $T(0) > 0; T(1) > 0; T(2) > 0; \ldots T(n-1) > 0$. In adding them all a great deal of cancellation occurs and we are left with the inequality $t(n) - t(0) > 7.5n$. This contradicts the original assumption that $t(n) - t(0) = 7.5n$, and so we can say that she *does* cover a continuous mile in exactly 7.5 minutes.

It turns out that this is true for only an integral number of miles, and that seems, frankly, rather strange. We'll examine this case for a specific function, by redefining $t(x)$ to be:

$$t(x) = \varepsilon \sin\frac{\pi x}{m} + 7.5x, \tag{6.1}$$

where *m* is *not* an integer and $\varepsilon > 0$. If ε is small enough and *m* is large enough, this will represent an increasing function with small "undulations" about the line 7.5*x*, representing the time to jog at the average speed. $t(x)$ is an increasing function if $t'(x) > 0$, as it must be because it is a time function. Mathematically this will be so if

$$\frac{\pi\varepsilon}{m}\cos\left(\frac{\pi x}{m}\right) + 7.5 > 0, \text{ i.e. if } \frac{\varepsilon}{m} < \frac{7.5}{\pi} \approx 2.387$$

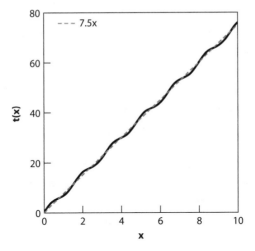

Figure 6.1. Lindsay's time function $t(x)$.

for this particular model of the jogging time. Figure 6.1 shows $t(x)$ and the mean time function for the case $\varepsilon = 1$, $m = 0.8$.

Now it follows that

$$t(x+1) - t(x) = \varepsilon\left(\sin\frac{\pi(x+1)}{m} - \sin\frac{\pi x}{m}\right) + 7.5$$

$$= 2\varepsilon \cos\left[\frac{\pi\left(x+\frac{1}{2}\right)}{m}\right]\sin\frac{\pi}{m} + 7.5$$

is an increasing function that can *never* be 7.5. Therefore if Lindsay runs so that her time function is given by equation (6.1), she will never run a complete mile in exactly 7.5 minutes.

$X = T$: RAINING IN THE CITY

It seems that it's almost impossible to find a taxi when it's raining in the city. Gene Kelly preferred to sing and dance, but, as sometimes happens when I am walking, the heavens open and I am faced with the different options of running fast or walking for shelter. Is it better to walk or *run* in the rain? The decision to run may seem an obvious one, but depending on several factors discussed below, it is not always that simple. And occasionally I have the foresight to take an umbrell.

What might such factors be? Among them are how fast the rain is falling, and in what direction (i.e., is there wind, and if so, in what direction), how fast I can run or walk, and how far away the shelter is. Let's ignore the wind and rain direction initially and set up a very basic model. Suppose that you run at v m/s. Since 2 mph $\approx$ 1 m/s, as is easily shown, we can easily convert speeds in mph to MKS units and vice versa. Of course, this is only approximate, but we shall nevertheless use it for simplicity; we do not require precise answers here; after all, we're in a hurry to get home and dry off!

Suppose the distance to the nearest shelter point is d km, and that the rain is falling at a rate of h cm/hr (or $h/3600$ cm/s). Clearly, in this simplified case it is better to run than to walk. Here is a standard classification of the rate of precipitation:

- Light rain—when the precipitation rate is < 2.5 mm ($\approx$ 0.1 in) per hour;

- Moderate rain—when the precipitation rate is between 2.5 mm ($\approx$ 0.1 in) and 7.6 mm (0.30 in) to 10 mm ($\approx$ 0.4 in) per hour;

- Heavy rain—when the precipitation rate is between 10 mm (0.4 in.) and 50 millimeters (2.0 in) per hour;

- Violent rain—when the precipitation rate is > 50 mm (2.0 in) per hour.

Now if I run the whole distance d m at v m/s, the time taken to reach the shelter is d/v seconds, and in this time the amount H of rain that has fallen is given by the expression $H = hd/3600v$ cm. If I run at 6 m/s (about 12 mph), for example, and $d = 500$ m, then for heavy rain (e.g., $h = 2$ cm/hr), $H \approx 0.5$ cm. This may not seem like a lot, but remember, it is falling on and being absorbed (to some extent at least) by our clothing, unless we are wearing rain gear. The next step in the model is to estimate the human surface area; a common approach is to model ourselves as a rectangular block, but a quicker method is to consider ourselves to be a flat sheet 2 m high and 0.5 m wide. The front and back surface area is 2 m² — about right! Over the course of my run for shelter, if all that rain is absorbed, I will have collected an amount $2 \times 10^4 \times 5 \times 10^{-2} = 10^3$, or one liter of rain. That's a wine bottle of rain that the sky has emptied on you! I prefer the real stuff...

It is important to note that rain is a "stream" of discrete droplets, not a continuous flow. It is reasonable to define a measure of rain intensity by comparing the rate at which rain is falling with the speed of the rain. The speed of raindrops depends on their size. At sea level, a very large raindrop about 5 millimeters across falls at the rate of about 9 m/s (see Chapter 25 for an unusual way to estimate the speed of raindrops). Drizzle drops (less than 0.5 mm across) fall at about 2 meters per second. We shall use 5 m/s (or $500 \times 3600 = 1.8 \times 10^6$ cm/hr) as an average value. The ratio of the precipitation rate to the rain speed [13] is called the rain intensity, I. For the figures used here, $I = 2/(1.8 \times 10^6) \approx 1.1 \times 10^{-6}$. Therefore I is a parameter: $I = 1$ corresponds to continuous flow (at that speed), whereas $I = 0$ means the rain has stopped, of course!

If as in Figure 6.2 the rain is falling with speed c m/s at an angle θ to the vertical direction, and you are running into it, the vertical (downward) component of speed is $c \cos \theta$ and the relative speed of the horizontal component is $c \sin \theta + v$. And if you move fast enough, only the top and front of you will get wet. For now let's assume this is the case. Since you are *not* in fact a thin sheet (your head does get wet), we will model your shape as a rectangular box of height l, width w and thickness t (all in m). The "top" surface area is wt m^2, and the volume of rain is "collected" at a rate $R =$ intensity $\times$ surface area $\times$ rain speed $= Iwtc \cos \theta$, expressed in units of m^3/s. In time d/v the amount collected is then $Iwtdc \cos \theta/v$ m^3. A similar argument for the front surface area gives $Iwld(c \sin \theta + v)/v$ m^3, resulting in a total amount T of rain collected as

$$T = \frac{Iwd}{v} \left[ct \cos \theta + l(c \sin \theta + v) \right] \equiv \frac{\alpha \beta}{v} + l;$$
$$\text{where } \alpha = Iwd; \ \beta = c(t \cos \theta + l \sin \theta). \tag{6.2}$$

Before putting some numbers into this, note that if θ were the only variable in this expression, then

$$\frac{dT}{d\theta} \propto (l \cos \theta - t \sin \theta).$$

Because this vanishes at $\theta = \arctan(l/t)$ and $d^2T/d\theta^2 < 0$ there, this represents a *maximum* accumulation of rain. Physically this means that you are running almost directly into the rain, but the relative areas of your top and front are such that the maximum accumulation occurs when it is (in this case) not quite

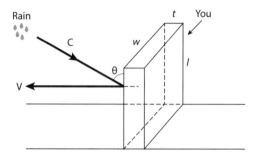

Figure 6.2. Configuration for a rectangular box-person running in the rain.

horizontal. In addition, if v were the only variable, then $dT/dv = -\alpha\beta/v^2 < 0$ if $\beta > 0$, so that T decreases with speed v if $\tan\theta > -t/l$. If $\tan\theta < -t/l$, then T *increases* with speed. This corresponds to more and more negative values of θ, that is, the rain is coming from behind the runner.

Let's put some meat on these bones, so to speak. Suppose that the rectangular box has dimensions $l = 1.5$ m, $w = 0.5$ m, and $t = 0.2$ m. Furthermore, we have chosen $v = 6$ m/s (about 13 mph); and $c = 5$ m/s. Substituting all these values into expression (6.2) for T we obtain $T \approx 4.6 \times 10^{-2}(9 + \cos\theta + 7.5\sin\theta)$ liters. It is readily confirmed that this has a maximum value of about 0.76 liters when $\theta \approx 82.4°$. A graph of $T(\theta)$ is shown in Figure 6.3 for $-\pi/2 < \theta < \pi/2$. Note also that for a given speed when $\theta < 0$ (i.e., the rain is coming from behind you) the quantity of rain accumulated is smaller (as would be expected) than when you are running into the rain. Notice also the wide range of values, from a high of just below 0.8 liters to a low of less than 0.1 liters when the rain is hitting you horizontally from behind!

We need to be a little more careful with the case of $\theta < 0$ because (6.2) could become negative (and therefore meaningless) for some parameter ranges. If we replace θ by $-\phi$, $\phi > 0$, then this equation becomes

$$T = \frac{lwd}{v}\left[ct\cos\phi - l(c\sin\phi - v)\right] \tag{6.3}$$

This is negative if

$$\sin\phi > \frac{t}{l}\cos\phi + \frac{v}{c}. \tag{6.4}$$

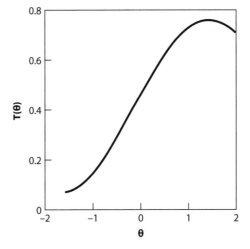

Figure 6.3. Total amount of rain captured as a function of rain angle (radians).

Note that this can never happen if $v \geq c$, and will not for the choice of v and c made here $(v/c = 1.2)$. If the inequality (6.4) *is* satisfied, the "offending term" comes from the "front accumulation" value

$$Iwld\,(v - c\sin\phi)/v,$$

which should now be written as

$$Iwld\,(c\sin\phi - v)/v$$

because the rain falls on your back if $v < c\sin\phi$. The correct total is now

$$T = \frac{Iwd}{v}\left[ct\cos\phi + l(c\sin\phi - v)\right]. \tag{6.5}$$

It is instructive to rewrite this equation as

$$T = Iwd\left[1 + \frac{ct\cos\phi}{v}\left(1 - \frac{l\tan\phi}{t}\right)\right]. \tag{6.6}$$

Let's focus our attention on the term in parentheses in equation (6.6), noting that $t/l = A_{top}/A_{back}$ is just the ratio of the top area of the human "box" to the area of the back (or front). If $\tan\phi > t/l = A_{top}/A_{back}$, this term is negative, and in this case, you should attempt to go no faster than the horizontal speed of the rain ($c\sin\phi$) at your back. Using equation (6.5) we see that if your speed increases so that $v = c\sin\phi$, you are just keeping up with the rain and T is minimized. This may seem at first surprising since for $v > c\sin\phi$, T is reduced still farther, but *now* you are catching up to the rain ahead of you, and it falls once more on your front (and head, of course). In this case formula (6.2) again applies.

How about putting some numbers into these formulae? For a generic height $l = 175$ cm (about 5 ft 9 in), shoulder to shoulder width $w = 45$ cm (about 18 in), and chest to back width $t = 25$ cm (about 10 in), the ratio $t/l = 1/7$, and so if $\tan\phi > 1/7$, that is, $\phi \approx 8°$, the ratio $v/c = \sin\phi \approx 1/7$ (see why?). Therefore if it is raining heavily at about 5 m/s from this small angle to the vertical, you need only amble at less than one m/s (about 2 mph) to minimize your accumulated wetness! Although the chosen value for w was not used, we shall do so now. The top area of our human box is ≈ 1100 cm^2, one side area is ≈ 4400 cm^2, and the front or back area is ≈ 7900 cm^2.

Exercise: Calculate these areas in square feet if you feel so inclined.

To summarize our results, if the rain is driving into you from the front, run as fast as you safely can. On the other hand, if the rain is coming from behind you, and you can keep pace with its horizontal speed by walking, do so! If you exceed that speed, the advantage of getting to your destination more quickly is outweighed by the extra rain that hits you from the front, since you are now running into it! Perhaps the moral of this is that we should always run such that the rain is coming from behind us!

$X = \Delta T$: WEATHER IN THE CITY

To some extent cities can create their own weather. No doubt you have heard of the sidewalk in some city being hot enough to fry an egg; include all the paved surfaces and buildings in a city, and you have the capacity to cook a lot of breakfasts! Typically, such surfaces get hotter than those in rural environments because they absorb more solar heat (and therefore reflect less), and retain that heat for longer than their rural counterparts by virtue of their greater thermal "capacity." The contrast between a city and the surrounding countryside is further enhanced at night, because the latter loses more heat by evaporative and other processes. Furthermore, the combined effects of traffic and industrial plants are a considerable source of heat within an urban metropolitan area. Thus there are several factors to take into account when considering local climate in a city versus that in the countryside. They include the fact that (i) there are differences between surface materials in the city and the countryside—concrete, tarmac, soil, and vegetation; (ii) the city "landscape"—roofs, walls, sidewalks, and roads—is much more varied than that in the country in the shape and orientation of reflective surfaces; vertical walls tend to reflect solar radiation downward instead of skyward (see Figure 6.4a,b), and concrete retains heat longer than do soil and vegetation; (iii) cities are superb generators of heat, particularly in the winter months; (iv) cities dispose of precipitation in very different ways, via drains, sewers, and snowplows (in the north). In the country, water and snow are more readily available for evaporative cooling.

Such local climate enhancement has several consequences, some of which are positive (or at least appear to be). For example, there may be a diminution of snowfall and reduced winter season in the city. This induces an earlier spring, other species of birds and insects may take up residence, and longer-lasting higher temperature heat waves can occur in summer (quite apart from any effects of larger scale climate change). This in turn means that less domestic heating may be required in the winter months, but more air-conditioning in the summer. The effects of an urban-industrial complex on weather generally are

(a)   (b)

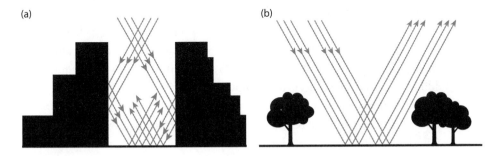

Figure 6.4. (a) Vertical surfaces tend to reflect solar radiation toward the ground and other vertical surfaces (thus trapping it), especially when the sun's elevation is moderately high. (b) There being fewer vertical surfaces in the countryside, solar radiation tends to be reflected skyward. Redrawn from Lowry (1967).

harder to quantify, though stronger convective updrafts (and hence intensity of precipitation and storms) are to be expected downwind from urban areas. According to one report (Atkinson 1968), there has been a steadily increasing frequency of thunderstorm activity near London as it has grown in size. In U.S. cities, the incidence of thunderstorms is 10–42% greater than in rural areas, rainfall is 9–27% greater and hailstorms occur more frequently, by an enormous range: 67–430%.

If the air temperature were to be recorded as we move across the countryside toward a city, the rural/urban boundary will typically exhibit a sharp rise—a "cliff"—followed by a slower rate of increase (or even a plateau) until a more pronounced "peak" appears over the city center. If the temperature difference between the city and surrounding countryside at any given time is denoted by ΔT, the average annual value for ΔT ranges from 0.6 to 1.8°C. Of course, the detailed temperature profile as a function of position will vary depending on the time of day, but generally this is a typical shape: a warm "island" surrounded by a cooler "sea." Obviously the presence of parks and other open areas, lakes, and commercial, industrial and heavily populated areas will modify this profile on a smaller spatial scale. The difference ΔT between the maximum urban temperature and the background rural temperature is called the *urban heat island intensity*. Not surprisingly, this exhibits a diurnal variation; it is at a maximum a few hours after sunset, and a minimum around the middle of the day. In some cases at midday the city is cooler than its environs, that is, $\Delta T < 0$.

To see why this might be so, note that near midday the sunlight strikes both country and city environs quite directly, so ΔT can be small, even negative, possibly because of the slight cooling effect of shadows cast by tall buildings, even with the sun high overhead. As the day wears on and the sun gets lower, the solar radiation strikes the countryside at progressively lower angles, and much is accordingly reflected. However, even though the shadows cast by tall buildings in the city are longer than at midday, the sides facing the sun obviously intercept sunlight quite directly, contributing to an increase in temperature, just as in the hours well before noon, and ΔT increases once more.

Cities contribute to the "roughness" of the urban landscape, not unlike the effect of woods and rocky terrain in rural areas. Tall buildings provide considerable "drag" on the air flowing over and around them, and consequently tend to reduce the average wind speed compared with rural areas, though they create more turbulence (see Chapter 3). It has been found that for light winds, wind speeds are greater in inner-city regions than outside, but this effect is reversed when the winds are strong. A further effect is that after sunset, when ΔT is largest, "country breezes"—inflows of cool air toward the higher temperature regions—are produced. Unfortunately, such breezes transport pollutants into the city center, and this is especially problematical during periods of smog.

Question: Why is ΔT largest following sunset?

This is because of the difference between the rural and urban cooling rates. The countryside cools faster than the city during this period, at least for a few hours, and then the rates tend to be about the same, and ΔT is approximately constant until after sunrise, when it decreases even more as the rural area heats up faster than the city. Again, however, this behavior is affected by changes in the prevailing weather: wind speed, cloud cover, rainfall, and so on. ΔT is greatest for weak winds and cloudless skies; clouds, for example, tend to reduce losses by radiation. If there is no cloud cover, one study found that near sunset $\Delta T \propto w^{-1/2}$, where w is the regional wind speed at a height of ten meters (see equation (6.7)).

Question: Does ΔT depend on the population size?

This has a short answer: yes. For a population N, in the study mentioned above (including the effect of wind speed), it was found that

$$\Delta T \approx \frac{N^{1/4}}{4w^{1/2}}, \tag{6.7}$$

though other studies suggest that the data are best described by a logarithmic dependence of ΔT on $\log_{10} N$. While every equation (even an approximate one) tells a story, equation (6.7) doesn't tell us much! ΔT is weakly dependent on the size of the population; according to this expression, for a given wind speed w a population increase by a factor of sixteen will only double ΔT! And if there is no wind? Clearly, the equation is not valid in this case; it is an empirical result based on the available data and valid only for ranges of N and w.

Exercise: "Play" with suitably modified graphs of $N^{1/4}$ and $\log_{10} N$ to see why data might be reasonably well fitted by either graph.

The reader will have noted that there is not much mathematics thus far in this subsection. As one might imagine, the scientific papers on this topic are heavily data-driven. While this is not in the least surprising, one consequence is that it is not always a simple task to extract a straightforward underlying mathematical model for the subject. However, for the reader who wishes to read a mathematically more sophisticated model of convection effects associated with urban heat islands, the paper by Olfe and Lee (1971) is well worth examining. Indeed, the interested reader is encouraged to consult the other articles listed in the references for some of the background to the research in this field.

To give just a "taste" of the paper by Olfe and Lee, one of the governing equations will be pulled out of the air, so to speak. Generally, I don't like to do this, because everyone has the right to see where the equations come from, but in this case the derivation would take us too far afield. The model is two-dimensional (that is, there is no y-dependence), with x and z being the horizontal and vertical axes; the dependent variable θ is essentially the quantity ΔT above, assumed to be small enough to neglect its square and higher powers. The parameter γ depends on several constants including gravity and air flow speed, and is related to the Reynolds number discussed in Chapter 3. The non-dimensional equation for $\theta(x, z)$ is

$$\left[\frac{\partial^2}{\partial z^2}\left(\frac{\partial}{\partial x} - \frac{\partial^2}{\partial z^2}\right) + \frac{\gamma}{4}\frac{\partial}{\partial x}\right]\theta = 0. \tag{6.8}$$

The basic method is to seek elementary solutions of the form

$$\theta(x, z) = \mathrm{Re}\left[\exp(\sigma z + ikx)\right]. \tag{6.9}$$

where "Re" means that the real part of the complex function is taken, and k is a real quantity On substituting this into equation (6.8) the following complex biquadratic polynomial is obtained:

$$\sigma^4 - ik\sigma^2 - i\left(\frac{\gamma k}{4}\right) = 0. \tag{6.10}$$

There are four solutions to this equation, namely,

$$\sigma = \pm\left(\frac{1+i}{2}\right)k^{1/2}\left[1 \pm \sqrt{1 - i\frac{\gamma}{k}}\right]^{1/2}, \tag{6.11}$$

but for physical reasons we require only those solutions that tend to zero as $z \to \infty$. The two satisfying this condition are those for which $\mathrm{Re}\,\sigma < 0$,

$$\sigma_1 = -\left(\frac{1+i}{2}\right)k^{1/2}\left[1 + \sqrt{1 - i\frac{\gamma}{k}}\,\right]^{1/2} \text{ and}$$

$$\sigma_2 = \left(\frac{1+i}{2}\right)k^{1/2}\left[1 - \sqrt{1 - i\frac{\gamma}{k}}\,\right]^{1/2}. \tag{6.12}$$

Using these roots, the temperature solution (6.9) can be expressed in terms of (complex) constants c_1 and c_2 as

$$\theta(x,z) = \mathrm{Re}\{[c_1 \exp(\sigma_1 z) + c_2 \exp(\sigma_2 z)]\exp(ikx)\}. \tag{6.13}$$

Note from equation (6.12) that $\sigma_1^2 = ik - \sigma_2^2$. Using specified boundary conditions, both c_1 and c_2 can be expressed entirely in terms of σ_1 and σ_2, though we shall not do so here. The authors note typical magnitudes for the parameters describing the heat island of a large city (based on data for New York City). The diameter of the heat island is about 20 km, with a surface value for $\Delta T \approx 2C$, and a mean wind speed of 3 m/s.

Exercise: Verify the results (6.10)–(6.12).

Chapter 7

NOT DRIVING IN THE CITY!

As we have remarked already, cities come in many shapes and sizes. In many large cities such as London and New York, the public transportation system is so good that one can get easily from almost anywhere to anywhere else in the city without using a car. Indeed, under these circumstances a car can be something of an encumbrance, especially if one lives in a restricted parking zone. So for this chapter we'll travel by bus, subway, train, or quite possibly, rickshaw. Whichever we use, the discussion will be kept quite general. But first we examine a situation that can be more frustrating than amusing if you are the one waiting for the bus.

$X = T$: BUNCHING IN THE CITY

In a delightful book entitled *Why Do Buses Come in Threes?* [8] the authors suggest that in fact, despite the popular saying, buses are more likely to come in twos. Here's why: even if buses leave the terminal at regular intervals, passengers waiting at the bus stops tend to have arrived randomly in time. Therefore an arriving bus may have (i) very few passengers to pick up, and little time is lost, and it's on its way to the next stop, or (ii) quite a lot of passengers to board. In the latter case, time may be lost, and the next bus to leave the terminal may have caught up somewhat on this one. Furthermore, by the time it reaches the next stop there may be fewer passengers in view of the group that boarded the previous bus, so it loses little time and moves on. For the next two buses, the cycle may well repeat; this increases the likelihood that buses will tend to bunch in twos, not threes.

The authors note further that *if* a group of three buses occurs at all (and surely it sometimes must), it is most likely to do so near the end of a long bus route, or if the buses start their journeys close together. So let's suppose that they do... that they leave the terminal every T minutes, and that once they "get their buses in a bunch" buses A and B and B and C are separated by t minutes (where $t < T$ of course). A fourth bus leaves T minutes after C, and so on. The four buses have a total of $3T$ minutes between them, initially at least. When the first three are bunched up, the fourth bus is $3T - 2t$ minutes behind C (other things such as speed and traffic conditions being equal). If you just missed the first or the second bus, you have a wait of t minutes for the third one; if you just missed that then you have a wait of $3T - 2t$ minutes. Thus your average wait time under these circumstances is just T, the original gap between successive buses!

How does this compare with *not* missing the bus? The probability that you have arrived in the long gap as opposed to one of the two short ones is $(3T - 2t)/3T$, your time of arrival could be just after the third bus left, just before the fourth one arrives, or any time in between, and the mean of those two extremes is $(0 + 3T - 2t)/2$. This will actually be longer than the previous average wait time if $T > 2t$, which of course is quite likely, even allowing for the small possibility that you arrived in a t-minute gap.

$X = V_a$: AVERAGE SPEED IN THE CITY

We shall examine and "unpack" an idealized analysis of public transport systems (based on an article [14] published in the transportation literature). The article contained a comprehensive account of two models: a corridor system and a network system. In so doing, it is possible to derive some general results for transport systems. Although some of these conclusions were represented in graphical form only, the mathematics behind those graphs is well worth unfolding here. In both models there are three components to the travel time from origin A to destination B: (i) the walk times to and from the station at either end; (ii) the time waiting for the transportation to arrive; and (iii) the total time taken by the subway train, bus, or train to go from the station nearest $A(S_A)$ to that nearest $B(S_B)$. (This includes waiting time at intermediate stops along the way.)

In what follows, the city is assumed to have a "grid" system with roads running N-S and E-W (see Appendix 3 on "Taxicab geometry"). While this may be more appropriate to cities in the United States, it is also an acceptable approximation for many European cities [14], [15]. From Figure 7.1 we see that a walk from any location to any station will involve an N-S piece and an E-W piece. If each station is considered to be at the center of a square of side L,

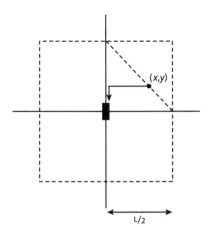

Figure 7.1. One segment of a transportation corridor with a central station serving a square area of side L. The diagonal shown has equation $x + y = L/2$.

L/2

the average walk length is $L/2$. To see this, note that for any trip starting at the point (x,y) on the diagonal line (for which $x + y = L/2$) to the center is just $L/2$. By symmetry, there is the same area on each side of the line. If the population demand is uniformly distributed, $L/2$ is the average trip to or from the central station, and hence L is the average total distance walked—just the average stop separation.

First we consider the average speed V_a for the vehicular part of the trip. We suppose that this part consists of a uniform acceleration a from rest to a (specified) maximum speed V_m (A–B), and a period of travel at this speed (B–C–D) followed by a uniform deceleration $-a$ to rest at the next station (D–E). The acceleration and deceleration take place over a distance x km, say, and the speed is constant for a distance $L - 2x \geq 0$. Using the equations of motion for uniform acceleration it is found that $V_m = (2ax)^{1/2}$ and the time T_x to travel the distance x ($=$ distance AB) and reach this maximum speed is $T_x = (2x/a)^{1/2}$. The time to travel the distance BC is therefore $T_{BC} = ([L/2] - x)/V_m$. If we include a waiting time T_W at the station A before leaving for the next one at E, the total time for the journey AE is $2(T_x + T_{BC}) + T_W$; typically, $T_W \approx 20$s and $a = 0.1g$ according to published data [15]. Hence the average speed in terms of the maximum speed is

$$V_a = \frac{L}{2(T_x + T_{BC}) + T_W} = \frac{L}{(V_m/a) + (L/V_m) + T_W} \qquad (7.1)$$

after a little reduction. We take $T_W = 20$s and $a = 10^{-3}$ km/s$^2 \approx 1.3 \times 10^4$ km/hr^2 in what follows. Then equation (7.1) takes the form

$$V_a = \frac{L}{\alpha V_m + L V_m^{-1} + \beta}, \qquad (7.2)$$

where $\alpha = 7.7 \times 10^{-5}$, $\beta = 5.6 \times 10^{-3}$.

It is interesting to note that on the basis of formula (7.2), V_a possesses a shallow maximum occurring at $V_m = (L/\alpha)^{1/2} \approx 114\sqrt{L}$ for the constants chosen here. For a station separation $L = 0.5$ km this corresponds to $V_m \approx 80$ km/hr. It can be seen from Figure 7.2 that for $L = 1$ km the maximum average speed attainable is only about 40 km/hr regardless of maximum speed. For a stop spacing of 0.25 km the maximum is less than 20 km/hr. In these cases the vehicle does not have enough time to reach its maximum speed and merely accelerates to the midpoint and then decelerates.

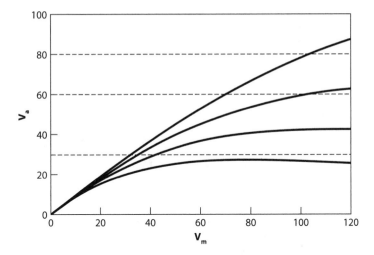

Figure 7.2. Average speed vs. maximum speed for different values of station spacing $L = 5$ km, 2 km, 1 km, 0.5 km.

The next phase of the calculation is to include passenger "walk and wait"times, the latter referring to the wait for the arrival of the train or bus. Furthermore, a trip length is now assumed. We shall adopt 80 m/min or 4.8 km/hr for the average walking speed, the average passenger wait time as 5 minutes, and an estimate of 8 km for the average (UK) trip length [14]. The number of stops beyond the origin is $8/L$ (if this is an integer), as is the number of in-station wait times T_W. For small values of L (less than 0.25 km, for example) the model is impractical, but we shall nevertheless treat L (and the number of stations) as a continuous variable. The average speed is now

$$V_a = \frac{L}{(8/L)[(V_m/a) + (L/V_m) + T_W] + (L/4.8) + (1/12)} \quad (7.3)$$

Graphs of $V_a(L)$ for $V_m = 30$ km/hr and 80 km/hr are shown in Figures 7.3a and 7.3b, for comparison with the graph for the in-vehicle average speed without the passenger walk-and-wait times (the above expression without the last two terms in the denominator). The lower speed for V_m is representative of a bus, given the typical stops necessary at crosswalks (pedestrian crossings) and traffic lights. The higher speed is more typical of light rail, monorail, or possibly priority bus lanes, which would ensure higher average speed. From the

(a)

(b)

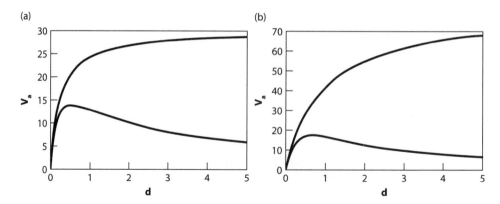

Figure 7.3. (a) Average speed vs. stop separation (in km) for maximum speeds of 30 km/hr for in-vehicle with stops (upper curve) and with walk and wait times included (lower curve). (b) Average speed vs. stop separation (in km) for maximum speeds of 80 km/hr for in-vehicle with stops (upper curve) and also with walk and wait times included (lower curve). The generic distance d in Figures 7.3–7.5 represents the spacing L between stops.

figure it is seen that for the bus the optimum spacing L between stations (corresponding to the peak average speed) is about 0.5 km. For the higher speed, the optimum spacing is closer to 0.75 km.

Note what little difference the increase in maximum speed makes to the maximum average speed. These average speeds are probably at best comparable to the average speed for cars in rush hour periods, since the latter will not have to keep stopping to let passengers on or off. This model suggests that overall trip times using public transport will generally be higher than for cars during the same peak times.

$X = L_v$: AVERAGE TRIP LENGTH IN THE CITY

Consider first an eastbound trip consisting of N stops, numbered from $n = 0$ to $N - 1$. The total number of such trips from 0 is $N - 1$, from 1 is $N - 2$, etc. so summing over all values of n we find that the total number of eastbound trips is the well-known result

$$\sum_{1}^{N-1} n = \frac{1}{2}N(N-1).$$

The sum of all the trips from the *m*th stop eastward ($0 \le m \le N-1$) can be found by replacing N above by $N-m$, that is,

$$\sum_{1}^{N-m-1} n = \frac{1}{2}(N-m)(N-m-1).$$

If the stations are a distance L apart, then the sum of *all* trip distances from *all* stops is L times the sum of the above expression over all possible values of *m*, that is,

$$L\sum_{m=0}^{N-1}\frac{1}{2}(N-m)(N-m-1) = \frac{N(N^2-1)L}{6}. \tag{7.4}$$

Exercise: establish the result (7.4).

For trips on a rectangular grid containing M stops on the *N-S* corridors, the total length of eastbound trips in any one row is M times the result (7.4) above, and the total length over all rows requires multiplication by M once more. Multiplying this by two for the westbound trips, the total length of eastbound and westbound trips is

$$L_{E,W} = \frac{M^2 N(N^2-1)L}{3}.$$

Assuming a uniform demand at the MN starting points, each of which serves the $MN-1$ others, there are $MN(MN-1)$ trips, so the average length of all eastbound and westbound trips is

$$L_{E,W}^a = \frac{M^2 N(N^2-1)L}{3MN(MN-1)} = \frac{M(N^2-1)L}{3(MN-1)}.$$

By interchanging M and N we obtain the corresponding result for N-S trips, that is,

$$L_{N,S}^a = \frac{N(M^2-1)L}{3(MN-1)}.$$

Any single trip will generally involve a combination of both types of trip, so the average length over all trips is

$$L_{av} = L^a_{E,W} + L^a_{N,S} = \frac{[M(N^2-1) + N(M^2-1)]L}{3(MN-1)} = \frac{(M+N)L}{3}. \quad (7.5)$$

This gratifying result shows that the average trip length is one sixth of the perimeter of the area served, and if the area is square, this is two thirds of the length of a side.

With this general result established (though only for the in-vehicle part of the trip), we proceed with the "walk, wait, ride, and walk" overall trip time calculations for a square city of side c. With a grid containing n^2 equally spaced stations, the distance between adjacent stops is c/n, and with each station centrally located in its "sub-square," the average walk distance is, from our earlier discussion, $c/2n$, which of course must be doubled for the assumed symmetry of the trip. The time to walk both such trips is c/nW.

The "wait time" depends on the spacing of the vehicles on a line. For example, if they are two stations apart the wait time is $2c/nV_a$, and in general, sc/nV_a for a vehicle spacing of s stations. The trip time is merely the average distance divided by V_a, and from the previous discussion, this will be $2c/3V_a$. There is also a transfer time for travel involving different vehicles, which for our purposes is just the waiting time multiplied by a factor F of order one.(There are data [14] indicating that for networks sizes up to 10×10, $0.5 \leq F \leq 2$. This will be generically incorporated in the term sc/nV_a for the calculations below.)

Combining all three "times," that is, walk + wait + in-vehicle, we express the overall trip time as

$$T = c\left[\frac{1}{nW} + \frac{s}{nV_a} + \frac{2}{3V_a}\right]. \quad (7.6)$$

In keeping with the literature [14], we choose a city size of $c = 10$ km. In Europe, typical population densities are around [14] 4000/ km^2 (the average figure for U.S. cities is about half this). Thus "our" city has a population of about 400,000. The average length of all trips is, as noted above, 2/3 of the side length, approximately 6.7 km. Again, fictionally treating the station spacing as a continuous variable, we can the network travel time as a function of station spacing from the above formula.

The total travel time in minutes is shown in Figure 7.4 for $V_m = 30$ km/ hr (upper curve) and 80 km/hr (middle curve). The lower (dotted) line is the

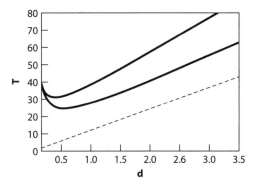

Figure 7.4. Total travel time (minutes) vs. stop spacing for a maximum speed of 30 km/hr (upper curve) and 80 km/hr (middle curve). The dotted line represents the walk time component for one end of the trip.

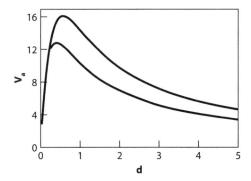

Figure 7.5. Average speeds in km/hr vs. stop spacing, based on Figure 7.4. The upper curve is for a maximum speed of 80 km/hr, the lower for a maximum speed of 30 km/hr.

walk time, and it can be seen that the walk time is the dominant contribution for large station spacing (corresponding to fewer tracks). Notice that the minimum travel time for the slower speed of 30 km/hr occurs at a stop spacing of about 0.25 km (appropriate for a city bus route) and at about 0.5 km for the 80 km/hr monorail route.

Figure 7.5 shows the average speed for these curves (starting for $d=L=$ 0.1 km). Note that the maximum average speed is only between about 13 and 16 km/hr for the range of maximum speeds considered here.

The much higher value for V_m of 80 km/hr adds relatively little to the average speed.

A final word: *pedestrians*. The most complete forms of the above models make allowances for the average time to walk to and from the station. As someone who has walked to work almost every weekday for twenty-eight years, I have learned to be very careful about crossing roads. I have to judge whether there is sufficient time to do so before the nearest vehicle reaches me. (One certainly hopes there is.) In fact, every individual has (in principle) a *critical time gap*, T say, above which crossing is acceptable and below which it is not. Mathematically, this can be represented by a step function

$$G(t) = 0, t \leq T,$$
$$G(t) = 1, t > T.$$

I think a step function is an entirely appropriate thing for pedestrians to have.

Chapter 8

DRIVING IN THE CITY

A s noted in the previous chapter, in many very large cities a car is not needed at all. But there also are many cities and towns where it is essential to have a vehicle. One advantage of driving over public transportation can be that one does not have to keep stopping at intermediate locations on the way to one's destination (traffic permitting). Of course, the daily commute can be extremely frustrating when it is of the stop-and-start variety, and the gas consumption becomes prohibitive. At the time of writing the price of petrol in the UK is far higher (by a factor of two) than the equivalent cost of gas in the U.S. It must be the different spelling that causes this. But before discussing that and other driving-related topics, let's start with an unrealistic but amusing and informative question.

$X = \bar{v}$: GASOLINE CONSUMPTION IN THE CITY

Question: A car travels from a home in the suburbs to a downtown office building, somehow maintaining a constant speed of 30 mph. On the return journey it maintains a constant speed of 60 mph. What is its average speed?

No, it's not 45 mph, sorry. The car spends twice as long traveling at 30 mph as it does returning at 60 mph, so the average speed will be "weighted"toward the lower speed. The correct answer is 40 mph. To see why, the average speed is defined as the total distance traveled divided by the total time (T) for the round trip. For those who are concerned that we have not specified the distance from the proverbial "A to B"—not to worry, let's just call it d. Then the average speed is

$$\bar{v} = \frac{2d}{T} = \frac{2d}{\dfrac{d}{30} + \dfrac{d}{60}} = 2\left(\frac{1}{30} + \frac{1}{60}\right)^{-1} = 40 \text{ mph.}$$

In fact, the above result is the *harmonic mean* of the two speeds. The harmonic mean of a set of numbers is the reciprocal of the arithmetic mean of the reciprocals! Put more obviously in mathematical terms, for a set of n numbers x_i, $i = 1, 2, 3, \dots n$, the harmonic mean H is

$$H = \left[\frac{1}{n}\sum_{i=1}^{n}\left(\frac{1}{x_i}\right)\right]^{-1}. \qquad (8.1)$$

That's the power of algebra for you! The incorrect answer of 45 mph is based on the arithmetic mean, which as we see, is not the appropriate measure of "average" for this problem. This difference is exacerbated by considering more than one "vehicle"; suppose that nine vehicles traverse a route one mile in length, and that all travel at the posted speed limit of 55 mph. I decide to walk the route at a reasonable pace, say 4 mph. The average speed for all ten trips, as computed by using the arithmetic mean is

$$\bar{v} = \frac{(9 \times 55) + (1 \times 4)}{10} = 49.9 \text{ mph.}$$

On the other hand, if we calculate the total distance traveled and divide it by the total time taken, the result is

$$\bar{v} = \frac{(9 \times 1) + 1}{\dfrac{1}{55} + \dfrac{1}{4}} = \frac{10 \times 4 \times 55}{59} = 37.3 \text{ mph},$$

a considerable difference.

Question: If we replace "speed" by "average speed" in the above question, does it change the result?

Exercise: Generalize the first result for speeds v_1 and v_2.

$X = \Delta E$: Question: How does gasoline consumption vary with speed?

In recent years, as well as several decades ago, the increasing costs of oil and gasoline have threatened to change the habits of American motorists, and in some cases has done so (albeit for a limited period of time, after which gas prices have declined, and everything reverts to the *status quo*). We know that at low speeds in low gears fuel consumption rate is relatively high because of lower efficiency in converting chemical energy to kinetic energy at these speeds. At high speeds the effects of air resistance (or drag) also increases the rate of fuel consumption. So it's certainly reasonable to conclude that there is an optimal speed (or range of speeds) for which the rate of fuel consumption is minimized.

The national 55-mph speed limit in the U.S. is no longer in existence, but in 1982 a newspaper article stated that one should "Observe the 55-mile-an-hour national highway speed limit." Furthermore, it stated that "For every five miles an hour over 50, there is a loss of one mile to the gallon. Insisting that drivers stay at the 55-mile-an-hour mark has cut fuel consumption 12 percent for [Company name] of Jacksonville, Florida—a savings of 631,000 gallons of fuel a year. The most fuel-efficient range for driving generally is considered to be between 35 and 45 miles an hour" ("Boost Fuel Economy," *Monterey Peninsula Herald*, May 16, 1982, as quoted in Giordano et al. 2003).

My first reaction on reading this was to wonder if the article is referring to fuel consumption for domestic as opposed to commercial vehicles, or both (which seems unlikely). And I very much doubt that the mpg losses are a linear function of speed above 55 mph as suggested. Surely there are more factors that affect how many miles per gallon we get from our vehicles, such as the age

of the engine (and maybe the driver!), the type of fuel used, the air temperature, the speed of travel, and the resistive forces of drag and road friction. Drag
will depend on the speed and shape of the vehicle and the prevailing atmospheric conditions at the ground (wind in particular). Friction will depend on
the condition of the tires, and the type and weight of the car. The nature of the
terrain is very important also: is it a flat, smooth road or an unpaved track in a
hilly or mountainous region? How good is the driver? Does he drive smoothly
where possible, or speed up/slow down in an irregular pattern, even if the road
conditions do not warrant it? Is she an experienced driver or a novice? These
and probably several other considerations all have bearing on the problem of
interest here.

A quick and dirty method

The drag force on most everyday objects is proportional to their cross-
sectional area A and the square of their speed v. Therefore driving at highway
speeds for a distance d will consume energy $E \propto Av^2d$. For a given value of d
and a small change in v (Δv), the corresponding percentage change in energy
expended is [16]

$$\frac{\Delta E}{E} = 2\frac{\Delta v}{v}. \tag{8.2}$$

While the change from driving at 70 mph (as many do) to 60 mph (about
14%) may not be considered "small," we'll use it anyway, and conclude that
there is a nearly 30% drop in fuel consumption. This makes a lot of sense!

$X = d_0$: TRAFFIC SIGNALS IN THE CITY

What thoughts typically run through your mind as you approach a traffic signal? Here are some likely ones: will it stay green long enough for me to continue through? Will it turn red in enough time for me to stop? What if it turns
yellow and someone is close behind me—should I try to stop or go through?
And perhaps related to the latter thought, "Is there a police car in the vicinity?"
　　Obviously we assume that the car is being driven at or below the legal speed.
If the light turns yellow as you approach the signal you have a choice to make:
to brake hard enough to stop before the intersection, or to accelerate (or coast)

and continue through the intersection legally before the light turns red. Unfortunately, many accidents are caused by drivers misjudging the latter (or going too fast) and running a red light.

The mathematics involved in describing the limits of legal maneuvers is straightforward: integration of Newton's second law of motion. Suppose that the width of the intersection is s ft and that at the start of the deceleration (or acceleration), time $t = 0$, and the vehicle is a distance d_0 from the intersection and traveling at speed v_0. If the duration of the yellow light is T seconds, and the *maximum* acceleration and deceleration are denoted by a_+ and $-a_-$ respectively, then we have all the initial information we need to find expressions for the two situations above: to stop or continue through. A suitable form of Newton's second law relates displacement x and acceleration a as

$$\frac{d^2 x}{dt^2} \left(= v \frac{dv}{dx} \right) = a, \tag{8.3}$$

from which follow the speed and displacement equations

$$\frac{dx}{dt} = at + v_0, v^2 = v_0^2 + 2ax, \text{ and } x = \frac{1}{2} at^2 + v_0 t.$$

We have chosen $x(0) = 0$. In order to stop before the intersection in at most T seconds, it follows from the second of these equations that

$$d_0 \geq \frac{v_0^2}{2a_-}. \tag{8.4}$$

On the other hand, to continue through, we note that the vehicle must travel a distance $d_0 + s$ in less than T, so that requiring $x(T) \geq d_0 + s$ in the third equation above yields the inequality

$$d_0 \leq v_0 T + \frac{1}{2} a_+ T^2 - s. \tag{8.5}$$

In each inequality, we have assumed that the maximum deceleration and acceleration are applied accordingly. Before treating the inequalities graphically, we write them in dimensionless form. This will reduce the number of parameters needed from five to four. To illustrate this, if we divide the distance d_0 by s, we obtain a dimensionless measure of distance in units of s, namely $\delta = d_0/s$. Similarly, we define dimensionless speed by $\sigma = v_0/(s/T)$, time by $\tau = t/T$, and acceleration by $\alpha_\pm = a_\pm/(s/T^2)$. Notice that τ is a measure of time in units of the yellow light duration. In these new units, the combined inequalities show that a driver may successfully choose either legal alternative provided that

$$\frac{\sigma^2}{2\alpha_-} \le \delta \le \sigma + \frac{\alpha_+}{2} - 1. \tag{8.6}$$

Now we are in a position to discuss reasonable ranges on these parameters, starting with the physical data. Typical ranges have been taken from the literature [17]. If we adopt the range for the speed approaching the intersection as 10 mph $\le v_0 \le 70$ mph (or approximately 15 fps $\le v_0 \le 100$ fps), with the bounds 20 ft $\le d_0 \le 600$ ft, 30 ft $\le s \le 100$ ft, 2 s $\le T \le 6$ s, and 3 ft/s$^2 \le |a_\pm| \le$ 10 ft/s^2, then we find that $0.3 \le \sigma \le 16.7$, $0.2 \le \delta \le 20$, and $0.12 \le \alpha_\pm \le 8.3$. Since we are measuring time in units of T, the results are relative, even though T itself can and does vary. In Figure 8.1 we plot a generic graph for the bounds on $\delta(\sigma)$ (based on equation (8.6)). There are four regions to consider. The region OAB corresponds to a domain with relatively low speed and small distance, and represents a portion of (σ, δ) space for which either option—stop or continue—is viable. Continuing, the region above ABC is a (theoretically infinite!) region with relatively low speed and large distance, so it is easy to stop before reaching the intersection. BCD is a region with relatively high speed and large distance, and represents a domain in which a violation (or accident) is likely. Finally, below OBD the region corresponds to relatively high speed and small distance, implying it is possible to continue through the intersection before the light turns red.

The various $\delta_i(\sigma)$ functions chosen here (with $i = 1, 2, 3, 4$, also based on equation (8.6)) are defined as follows:

$(i)\, \delta_1(\sigma) = \sigma + 3.15$; $(ii)\, \delta_2(\sigma) = \sigma^2/10$;
$(iii)\, \delta_3(\sigma) = \sigma - 0.94$; $(iv)\, \delta_4(\sigma) = \sigma^2/16.6$;
$(v)\, \delta_5(\sigma) = \sigma^2/0.24$.

As in Figure 8.1, Figure 8.2 shows dimensionless position/speed graphs identifying regions of safety. Obviously this is a simplistic analysis of the stoplight problem; an experienced and careful driver will have developed some measure of intuition (and caution) concerning whether a successful "continue through" is possible. We have not considered the possibility of skids; they are

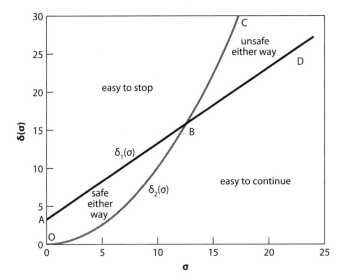

Figure 8.1. Dimensionless distance-speed graphs indicating regions of legal/illegal options based on equation (8.6); $\delta_1(\sigma) = \sigma + 3.15$ and $\delta_2(\sigma) = \sigma^2/10$.

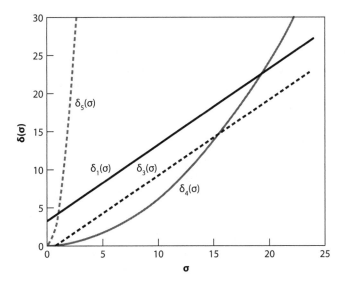

Figure 8.2. Additional dimensionless distance-speed graphs based on equation (8.6) (see Figure 8.1). Here δ_1 is as in Figure 8.1, and $\delta_3(\sigma) = \sigma - 0.94$; $\delta_4(\sigma) = \sigma^2/16.6$; and $\delta_5(\sigma) = \sigma^2/0.24$.

likely to occur if the deceleration is too large, and poor road conditions (wet, icy, etc.) will greatly affect the required stopping distance. In addressing this problem, Seifert (1962) has suggested posting signs along the roadside indicating a speed at which it is safe to continue through or stop from that location. It's a thought!

$X = \beta$: AVOIDING ACCIDENTS IN THE CITY

Should the driver of a car try to stop or turn in order to avoid a collision? We shall examine this question for several different situations, the first being when a car approaches a T-intersection with a brick wall directly ahead across the intersection. We shall assume that the junction is free of other vehicles, so the only possibility of collision involves the car hitting the wall. Furthermore, we shall assume that there is no skidding, in which case the coefficient of friction in a turn may be considered to be the same as that in the forward direction. (Skidding would involve the coefficient of *sliding* friction, in general different from that for rolling friction.)

We can examine three possible choices [18] as illustrated in Figure 8.3: (i) to steer straight ahead and apply the brakes for maximum deceleration; (ii) to turn in a circular arc without braking (using all the available force for centripetal acceleration); or (iii) to choose some combination of (i) or (ii), such as turning first and then steering straight (or vice versa), or even steering in a spiral path. In fact, option (ii) can be ruled out immediately by means of a simple (but nontrivial) argument as follows.

Suppose that the distance of the car from the wall is l and that its speed at that point is v_0. The force required to turn the vehicle (of mass m) in a circular arc of radius l is $F_1 = mv_0^2/l$, but the force required to bring the vehicle to stop in a distance l is $F_2 = mv_0^2/2l = F_1/2$. This means that if the car can be turned without hitting the wall, it can be brought to a stop halfway to the wall. Regarding option (iii), a rather more subtle argument [18] shows that the appropriate choice is still to stop in the direction of motion. Apart from a brief discussion below, this will not be elaborated on here; instead we shall examine some other potential driving hazards. It should be no surprise that the worst highway accidents are those involving head-on collisions.

A related problem is this: suppose that we are driving along a road in the right-hand lane for which, at the present speed, our stopping distance is D.

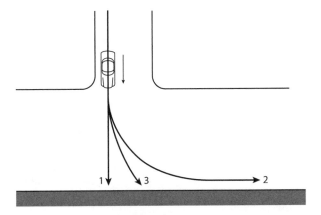

Figure 8.3. The three maneuver options open to the driver.

There is an obstacle ahead—it might be a repair crew, a stalled truck, or even a vehicle moving more slowly than we are (in the latter case we must adjust our speed in the calculations to that relative to the vehicle). What is the maximum obstacle width W that can just be avoided by turning left in a circular arc (if traffic in the adjacent lane permits this maneuver)?

We already know that the turning radius is twice the stopping distance. From Figure 8.4 we see that the angle $\beta = \alpha/2$. Here β is the angle subtended by the obstacle and the vehicle before the maneuver begins, and $\alpha = \arcsin$ $(1/2) = 30°$, so $\beta = 15°$. The width of the obstacle at this distance is therefore $D\tan\beta \approx 0.268\,D$. A wider obstacle cannot be avoided except by stopping.

Exercise: From the figure show that $\tan 15° = 2 - \sqrt{3}$.

Let's try to estimate some typical stopping distances for a range of speeds. This distance depends on the coefficient of friction (μ) between the tires and the road, and the driver's reaction time. The *minimum* such distance D_m can be found by ignoring the latter, as long as one adds the "reaction time × speed" distance D_r afterward. The frictional force must reduce the kinetic energy of the car to zero over the distance D_m. Provided the wheels do not lock during the deceleration (no sliding or skidding occurs), we use the coefficient of *static*

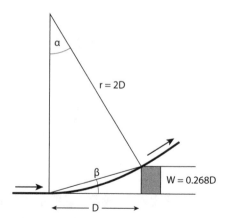

Figure 8.4. Geometry for turning to avoid a slower vehicle.

friction. If the wheels are locked, the braking force is due to *sliding* friction, which is in general different, as noted earlier. In the case of static friction, for a car of mass m, the equation to be satisfied is therefore

$$\frac{1}{2}mv_0^2 = \mu mgD_m,$$
(8.7)

from which it follows that

$$D_m = \frac{v_0^2}{2\mu g}.$$
(8.8)

Examining this result, we note that it is independent of the mass (or weight) of the car. It is also proportional to the square of the initial speed; thus doubling the speed *quadruples* the minimum stopping distance. The value of the coefficient μ depends on the quality of the tires and the prevailing road conditions; probably the best realistic value is $\mu = 0.8$, but for more worn tires, or wet roads, a somewhat lower value 0.7 or 0.6 is probably appropriate (or even lower for tires in poor condition). Here are some minimal (rounded) stopping distances for various speeds, taking $\mu = 0.65$. Also included, for illustrative purposes, is D_r, the distance covered in a nominal (and somewhat slow) reaction time of one second. The fourth column is the approximate total distance (D_T) required to stop at these speeds, given the above assumption.

v_0	D_m	D_r	D_T
20 mph (32 km/hr)	21 ft (6 m)	29 ft (9 m)	50 ft (15 m)
30 mph (48 km/hr)	46 ft (14 m)	44 ft (13 m)	90 ft (27 m)
50 mph (80 km/hr)	129 ft (39 m)	73 ft (22 m)	202 ft (61 m)
70 mph (113 km/hr)	252 ft (77 m)	103 ft (31 m)	355 ft (108 m)

For simplicity we now consider a minimal stopping distance of 100 ft, corresponding to a speed of about 44 mph (71 km/hr). If the vehicle is able to pass the obstacle without braking at all, this maneuver will begin after the reaction time. This means that the vehicle can pass a large obstacle of width nearly 27 feet if traffic in the adjacent lane(s) permits. But then the direction of the car will be at an angle of 30° to the original direction, a dangerous predicament to be sure!

To improve the safety of this maneuver, consider the following modification: we require that once abreast of the obstacle, the car should be moving parallel to the road in the new lane. For this case, the geometry changes a little (see Figure 8.5). Now the car's "trajectory" will be a sigmoidal-type shape

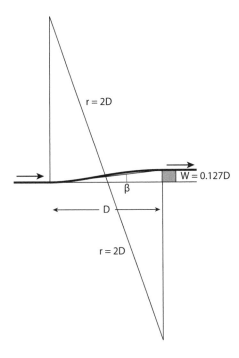

Figure 8.5. Modified geometry for turning to avoid a slower vehicle.

composed of two smoothly joined circular arcs as shown. As before $\beta = \alpha/2$, but now $\alpha = \arcsin(1/4) \approx 14.48°$, so $\beta \approx 7.24°$, meaning that the width of the obstacle should not exceed $D \tan \beta \approx 0.127 D$ for the maneuver to be executable. For a value of $D = 100$ ft, this is just less than 13 ft, which allows for a few feet of clearance around a large truck.

Next we consider two cars approaching an intersection perpendicularly at the same speed, as shown in Figure 8.6. Suppose that each driver instinctively tries to swerve to the side by at least $45°$; we will take this as a lower bound, for then they end up moving parallel to each other (if road conditions permit, of course). The angles of the truncated triangle are each $45°$ and therefore by symmetry the line joining each vertex to the corner of the junction makes an angle exactly half this with the hypotenuse. Recall that the minimum distance required for a "straight stop" is D and that for a circular arc is $2D$. Now the radius of the arc shown in Figure 8.6 is $r = D \cot(\pi/8) \approx 2.41D$. From equation (8.8) we can compare the corresponding speeds for the circular arc (1) and the $45°$ swerve (2):

$$\frac{D_m(2)}{D_m(1)} = \frac{v_0^2(2)}{v_0^2(1)} \approx 1.21.$$

This implies that $v_0(2)/v_0(1) \approx 1.1$, that is, the speed can be about 10% greater in the $45°$ swerve.

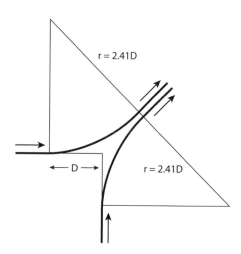

Figure 8.6. Two vehicles approaching on an initially perpendicular path.

A compromise of sorts between the straight stop of length D and the circular turn of radius $2D$ is the "spiral stop." This is accomplished by turning and braking simultaneously, and the vehicle will trace out a spiral path. The straight stop is achieved by applying (in the simple case) a constant force at $180°$ to the direction of motion. The circular path arises when that force is applied at $90°$ to the direction of motion. When this force is applied at some other constant angle γ to the direction of motion, the result is an equiangular spiral trajectory. In polar coordinates the equation of the path takes the form $r_1(\theta) = r_0 e^{(2\cot\gamma)\theta}$, where r_0 is a constant (see Appendix 6 for details).

$X = \sigma$: ACCELERATION "NOISE" IN THE CITY

What factors are important in studying what might be called "traffic engineering"? Clearly, weather and road conditions and the driver's response to a changing traffic environment are all significant and indeed, interrelated. A twisty two-lane road poses different problems from a six-lane Interstate highway or "main drag" in a city. The effects of increased traffic volume, road repairs, or adjacent building projects on congestion are generally difficult to answer quantitatively, but the concept of acceleration noise—the root-mean-square of the acceleration of a vehicle—is a useful one in determining some answers to questions of this type.

If $v(t)$ and $a(t)$ are respectively the speed and acceleration of a vehicle at time t, assumed continuous, then the average acceleration over a trip lasting time T is

$$\bar{a} = \frac{1}{T}\int_0^T a(t)\,dt = \frac{1}{T}[v(T) - v(0)]. \tag{8.9}$$

It is interesting to note that if the initial and final speeds are the same, then $\bar{a} = 0$. This will certainly be the case if the vehicle starts from rest and stops at the end of the trip!

The *acceleration noise* σ is the RMS (root-mean-square) of $a(t) - \bar{a}$, that is,

$$\sigma = \left[\frac{1}{T}\int_0^T [a(t) - \bar{a}]^2\,dt\right]^{1/2}. \tag{8.10}$$

Exercise: Show that

$$\sigma = \left[\frac{1}{T}\int_0^T [a(t)]^2 dt - \bar{a}^2\right]^{1/2}. \tag{8.11}$$

Clearly, when $\bar{a} = 0$, then

$$\sigma = \left[\frac{1}{T}\int_0^T [a(t)]^2 dt\right]^{1/2}. \tag{8.12}$$

Exercise: Calculate this quantity for several simple analytic choices of $a(t)$.

Why might this concept be a useful one for traffic engineering? If we think about a car that is driven fairly smoothly (i.e., with no violent acceleration or braking), we would expect the quantity σ to be small (in a sense to be discussed later). If the vehicle is driven with such accelerations and decelerations, σ will be large. Recall that the slope of a speed-time graph at a given point is the acceleration at that point. In effect, the acceleration noise is a measure of the smoothness of the speed-time graph—the smaller σ, the smoother the journey. A narrow, crowded road with sharp turns will give rise, other things being equal, to a higher value of σ. Those reckless drivers (never ourselves, of course) we see so frequently weaving in and out of traffic will engender high σ-values. Instead of wishing to shake a fist at them, or inwardly raging at them, perhaps we should just content ourselves with this fact. At this point, an amusing image comes to mind: after passengers alight from a car, they each in turn raise a card above their heads with an estimate of the σ-value for the just-completed trip!

In a very interesting article by Jones and Potts (1962), several possible answers to this question are provided, based on experimental data. After extensive investigations in Adelaide and its environs, they were able to draw several conclusions, some of them perhaps not surprising:

1. Given, say, a two-lane road and a four-lane road through hilly countryside, σ is much greater for the former than the latter.

2. For roads in hilly countryside, σ is smaller for an uphill journey than for a downhill one.

3. If two drivers drive at different speeds below the "design speed" of a highway, σ is about the same.

4. If one or both drivers exceeds the design speed of the highway, σ is higher for the faster driver.

5. An increase in the volume of traffic increases σ.

6. Similarly, an increase in traffic congestion resulting from parked cars, frequently stopping buses, cross traffic, pedestrians, etc., increases σ.

7. A suitably calibrated value of σ may provide a better measure of traffic congestion than travel or stopped times.

8. High values of σ indicate a potentially dangerous situation; they *may* be a measure of higher accident rates.

Naturally we must ask, what are typically high and low values for σ? The authors found that $\sigma = 1.5$ ft/s^2 is a high value, and $\sigma = 0.7$ ft/s^2 is a low one.

Another factor influenced by the size of σ is an economic one: fuel efficiency. This is not just important for trucking companies or individual truckers, of course: it is increasingly important for the average driver as well. Trucks are fitted with tachographs to record the driving behavior of the truckers, and presumably have the effect of providing an incentive to (in effect) lower the value of σ.

Generally, then, the smoothness of a journey can be measured by the acceleration noise—the standard deviation of the accelerations; furthermore, it is known that the acceleration distribution is essentially normal. It could well be a useful measure of the danger posed by erratic drivers, for whom σ is high. It also increases in tunnels, the reasons probably being narrow lanes, artificial lighting, and confined conditions. However, it is not always sensitive to changing traffic conditions, especially in city centers or major highways leading into them, where traffic congestion is common and the average speeds are low. It is also the case that a given value of σ may correspond to more than one traffic situation, for example, journeys at low speeds in dense traffic or faster journeys interspersed with traffic lights. One possibility to avoid this ambiguity is to use a modification of σ, namely $\tilde{\sigma} = \sigma / \bar{v}$, where $\bar{v}$ is the average speed. $\tilde{\sigma}$ may be interpreted as a measure of the mean change in speed per unit distance of the journey.

Question: *jerk* is defined as the time derivative of the acceleration, that is, $j(t) = da/dt$. Do you think there is any usefulness to defining a quantity similar acceleration noise (as in equation (8.10), namely *jerk noise*?

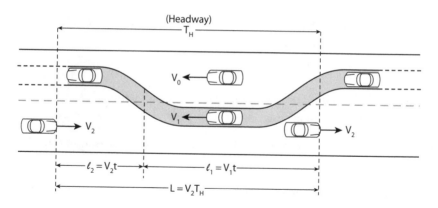

Figure 8.7. Headway geometry for a two-lane highway.

$X = T_h$: OVERTAKING IN THE CITY

We conclude this chapter with a simple mathematical model of overtaking and passing on a straight section of road. In Figure 8.7 a vehicle in the right lane moves into the left lane at speed $V_1 > V_0$ to overtake and pass a slower-moving vehicle. The equations, if not the diagram, are independent of which side of the Atlantic the vehicle resides! The *headway* T_H is the time gap between adjacent oncoming vehicles moving at speed V_2.

Clearly,

$$L = l_1 + l_2 = (V_1 + V_2)t = V_2 T_H, \text{ so } T_H = \left(1 + \frac{V_1}{V_2}\right)t. \quad (8.13)$$

Suppose that a reasonable time in which to complete a passing maneuver by a car traveling at 55 mph is 8 seconds, the headway required for an oncoming vehicle traveling at 65 mph in the opposing lane is $T_H \approx 15$ seconds. For interstates or other multilane highways, the needed gaps are smaller; we just change the sign of V_2 (obviously we require $V_1 < V_2$). In this case for the same speeds $T_H \approx 1.2$ seconds!

Chapter 9

PROBABILITY IN THE CITY

There are many topics under the umbrella "road traffic" that are amenable to mathematical investigation. Examples include traffic flow on the open road and at intersections, parking problems, accident rates, design of road systems for new towns and expanding cities, traffic lights and other control systems, transportation and scheduling problems, to name but a few. And as in so many aspects of mathematical modeling, there are two basic choices: to define the problem in a *deterministic* or a *probabilistic* context. The former can subdivide farther into continuum or discrete (or equivalently, macroscopic or microscopic) approaches, each with its own advantages and disadvantages. In the continuum approach the flow of traffic is treated as a fluid; properties of individual vehicles ("fluid particles") are not considered. Such models are often referred to as being *kinematic*—the motion (as opposed to the "forces" behind that motion) is of prime consideration. By contrast, car-following models are *dynamic* in the sense that accelerations and decelerations (and by implication, the forces) are implicit in their description. Although it is outside the scope of this book, it is perhaps worth mentioning that the discrete approach is often used to model flocking in starlings, shoaling in fish, and other swarming behavior (in ant, bee, and locust populations, for example).

One very interesting feature common to both natural and man-made phenomena is the topic of stability vs. instability. Will those waves on the lake grow or die out? Will that crack in the windshield continue to grow or stop? (Almost certainly the former!) Will that traffic bottleneck disappear by the time I get there, or will it get worse? A typical traffic "instability" arises from time lags in the response of a driver to the accelerations and decelerations of the vehicle in front. A small lag can grow as it passes, in a wavelike manner from car to car. Indeed, it is accurate to say that terms like "shock wave" and "expanding wave" are quite appropriate to describe some traffic patterns. Typically the existence of such waves in traffic flow can be deduced from both kinematic (continuous fluid-like) models and discrete ("particle") car-following models. Most are too detailed for inclusion here, but we can gain some insight into these phenomena by examining some simple "steady-state" models. In this context, the steady state represents an equilibrium or perhaps a neutrally stable solution, much like a ball on a flat table—if disturbed, it will not move away indefinitely (instability) or return to its starting point (stability) but will remain where it is placed. This is neutral stability. But we start with an introduction to the probabilistic approach to traffic flow. And as pedestrians, we should be particularly interested in the *gaps* between the vehicles! We shall think about the gaps in what follows.

$X = Pr$: PROBABILITY IN THE CITY

Probabilistic (or stochastic) models incorporate, by definition, an element of randomness. This word is not to be understood in the common sense as haphazard; a more precise definition is given below. In this context it can mean that there is a probability distribution for, say, the size of gaps in a line of traffic. We can view traffic or the gaps in traffic as a distribution in space or time. In space, a length of single lane road (for simplicity) will have a distribution of vehicles along it at any given moment in time—a snapshot view. Alternatively we may identify a fixed location on the road with vehicles passing this position as time goes on. The first situation is a distribution of intervals in space, and the second is a distribution in time. Such distributions (or series of events) are termed *random* provided that [19]

(i)   each event (e.g., vehicle arrival time) is independent of any other, and

(ii)   equal intervals of time (or space) are equally likely to contain equal numbers of events (e.g., vehicles).

There are several important distributions of interest in traffic flow studies; we shall briefly examine two of them—the *Poisson* and (displaced) *negative exponential distributions*. The latter is a simple generalization of a negative exponential distribution. The former gives the probability of a specified number of vehicles along a section of road at a given time, or passing a given point in a certain time interval. The latter provides the probability of a time or distance "gap" of a specified length in a specified time or distance. More precisely, it describes the time between events in a *Poisson process*, that is, a process in which events occur continuously and independently at a constant average rate. A derivation of the Poisson distribution is given in Appendix 4.

$X = P(t)$: TRAFFIC GAPS IN THE CITY?

This is an important question that has direct relevance to a topic mentioned above: pedestrians crossing roads. When I walk to work I have several roads to cross, and not all of them have crosswalks. There have also been quite a few occasions when I am in the middle of a crosswalk and cars go right by me (once it was a police cruiser that nearly knocked me down). There is a flavor of probability theory in this chapter, but not to worry, the applications we'll be making are very straightforward.

We'll call $P(t)$ the probability that no vehicle passes a certain point in a time interval t. Suppose that, over another period of time T a large number N of cars pass that same point. What do we mean by large in this context? Let's take $N > 100$. The average number of cars in a time interval t is then $n = Nt/T$. For events that are equally likely to occur at any time, the distribution of times between the events is well described *exponential distribution*. For example, it is often used for modeling the behavior of items with a constant failure rate.

It also has the advantage of taking a simple mathematical form. We define the exponential distribution

$$P(t) = e^{-n} = e^{-Nt/T}.$$ (9.1)

$P(t)$ also describes the probability that there is a gap of at least t seconds between the passage of any two consecutive vehicles. Therefore for N cars the average number of gaps $\geq t$ is given by

$$P_N(t) = NP(t) = Ne^{-Nt/T}.$$ (9.2)

An old survey [19] (see Gerlough 1955) of traffic gaps in the Pasadena (Arroyo Seco) Freeway over a period of 1753 seconds (!) in one lane found that 214 cars passed the observation point. By fitting a curve to the experimental data, that is, a plot of the number of gaps of length L or greater vs. the length of the gaps, the traffic researchers found quite good agreement between the data and the curve predicted by equation (9.2), namely

$$P_N(t) = 214e^{-214t/1753} = 214e^{-0.122t}.$$ (9.3)

X = N(W): GAPS AT SCHOOL CROSSINGS

Similar principles apply here, though a little preliminary work is necessary. What is the average walking speed of a school-aged child? Let's take the arithmetic mean of 1 mph (for the very young, or dawdlers!) and 3 mph for older children. Now

$$2 \text{ mph} = 2\frac{\text{miles}}{\text{hr}} \times \frac{5280 \text{ ft}}{1 \text{ mile}} \times \frac{1 \text{ hr}}{3600 \text{ s}} \approx 3 \text{ ft/s} \approx 1 \text{ m/s}$$

by the way; a result used in Chapter 6. If the width of the street in feet is W, then the time to cross the street is $t_c = W/3$ seconds. Of course, school crossings generally have guards who halt the traffic when the build-up of waiting children is sufficiently large. We shall dispense with the guards here and rely on natural gaps in the traffic to permit an opportunity to cross, say on average once every minute. We'll assume that the traffic flow past the crossing is governed by a distribution with a flow rate of N vehicles per hour. We wish to find the maximum flow rate, N_{max}, which would permit the above crossing opportunity. If $N > N_{max}$ then we'll reinstate the crossing guards!

The expected (or average) number of vehicles in the crossing t_c is $Nt_c/3600$, so as above, the probability that no vehicle will pass in that interval is

$$P(t) = e^{-Nt_c/3600}.$$

For k successive intervals of this length, on average kP of these will have no vehicular traffic, and for there to be one of these, $k = P^{-1}$. These k intervals correspond to a time $\tilde{t} = kt_c = t_c/P$. Therefore

$$t_c = \tilde{t}P = \tilde{t}e^{-Nt_c/3600},$$

and so

$$\ln(t_c/\tilde{t}) = -Nt_c/3600,$$

that is, if $\tilde{t} = 60s$ (corresponding to at least one crossing opportunity per minute),

$$N = N_{max} = \frac{10,800}{W}(5.19 - \ln W). \qquad (9.4)$$

A graph of this function is shown in Figure 9.1, drawn for a minimum width $W = 20$ ft. Notice how rapidly the maximum permissible flow rate decreases with the width of the street, particularly in the 20–40 ft range. It makes much sense then, for crossing guards to be in place especially in the vicinity of schools located near large highways.

We return to the question: how might the vehicles be distributed (in time) along the roadway? One possibility is the above-mentioned Poisson distribution (Chapter 3; see also Appendix 4), from which the probability of n arrivals in unit time is

$$P(n) = \frac{\lambda^n e^{-\lambda}}{n!}, \, n = 0,1,2,\dots \qquad (9.5)$$

Another possible model is the second one described above: the displaced exponential distribution with density function

$$f(t) = \lambda e^{-\lambda(t-a)}, \, t > a, \, a \geq 0. \qquad (9.6)$$

From this, the probability of encountering a gap exceeding a given time interval T is

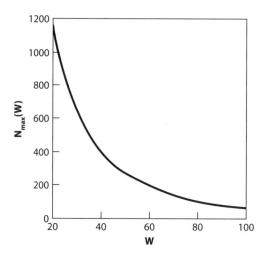

Figure 9.1. Maximum flow rate vs. width of road.

$$P(\text{gap} > T) = \int_{T}^{\infty} \lambda e^{-\lambda(t-a)} \, dt = e^{-\lambda(T-a)}.$$

The constant a is the minimum time gap (≈ 1 second for example). For a traffic flow rate of 600 vehicles per hour we can take $\lambda = 1/6$ vehicle/second. Suppose that a (fairly nimble) pedestrian requires a gap of at least $T = 6$ seconds to successfully cross the road; then with $a = 1$ the corresponding probability of being able to do so is $e^{-5/6} \approx 0.43$. Note also that

$$P(\text{gap} \leq T) = 1 - e^{-\lambda(T-a)}.$$

Increasing volume of traffic has obvious consequences. It generally causes a reduction in the mean speed of vehicles and also can affect the mean spacing between them. If this mean spacing is measured in terms of time rather than distance, as introduced at the end of Chapter 8, it is called a *headway*.

It's time for a little integration practice; let's combine it with some useful definitions. This will enable us to derive in a straightforward manner some results of interest in studies of pedestrian delays and minor-road delays to vehicles [19]. As is standard practice in probability theory, we define the expectation $E(x)$ for a continuous random variable taking on values in (b, c) with *probability density function* (p.d.f.) $f(x)$; it is

$$E(x) = \int_b^c x f(x)\, dx. \tag{9.7}$$

Note that by definition of a p.d.f., the integral $\int_b^c f(x)dx = 1$, so that for future reference we can write the expectation value as

$$E(x) = \frac{\int_b^c x f(x)\, dx}{\int_b^c f(x)\, dx}. \tag{9.8}$$

The expectation can be thought of as a mean value of the random variable. We shall consider the above displaced exponential distribution, which gives the distribution of lengths of intervals $\geq a$ between vehicles (headways). We can therefore calculate the mean headway time for all vehicles as

$$\int_a^\infty t\lambda e^{-\lambda(t-a)}\, dt = \left[-t e^{-\lambda(t-a)} \right]_a^\infty + \int_a^\infty e^{-\lambda(t-a)}\, dt = a + \frac{1}{\lambda}. \tag{9.9}$$

Obviously this reduces to λ^{-1} when $a = 0$, as it should (recall that λ is just the mean number of vehicles arriving in unit time). There are several other related properties of this distribution that are of interest to traffic engineers. The proportion of intervals in the interval (a, t) is

$$\int_a^t \lambda e^{-\lambda(\tau-a)}\, d\tau = 1 - e^{-\lambda(t-a)}.$$

The proportion of intervals $> t$ seconds is therefore $e^{-\lambda(t-a)}$. The proportion of *time* occupied by intervals $\leq t$ seconds is the weighted average

$$\frac{\int_a^t \tau\lambda e^{-\lambda(\tau-a)}\, d\tau}{\int_a^\infty \tau\lambda e^{-\lambda(\tau-a)}\, d\tau} = \frac{\left[(\tau+\lambda^{-1}) e^{-\lambda(\tau-a)} \right]_a^t}{\left[(\tau+\lambda^{-1}) e^{-\lambda(\tau-a)} \right]_a^\infty} = 1 - \left(\frac{1+\lambda t}{1+\lambda a} \right) e^{-\lambda(t-a)}. \tag{9.10}$$

Not surprisingly, the proportion of time occupied by intervals $> t$ seconds is

$$\left(\frac{1+\lambda t}{1+\lambda a} \right) e^{-\lambda(t-a)}.$$

Finally, the mean headway time for all intervals $\leq t$ seconds is given by the expression

$$\frac{\int_a^t \tau\lambda e^{-\lambda(\tau-a)}\, d\tau}{\int_a^t \lambda e^{-\lambda(\tau-a)}\, d\tau} = \frac{1}{\lambda} + \frac{a - t e^{-\lambda(t-a)}}{1 - t e^{-\lambda(t-a)}}. \tag{9.11}$$

The corresponding result for all intervals greater than t $(> a)$ seconds is

$$\frac{\int_{t}^{\infty} \tau \lambda e^{-\lambda(\tau-a)} d\tau}{\int_{t}^{\infty} \lambda e^{-\lambda(\tau-a)} d\tau} = t + \frac{1}{\lambda},$$

which is a result (not surprisingly) independent of a.

Exercise: Practice your integration by verifying equations (9.9)–(9.11).

Negative exponential distributions are sometimes considered even more important than the Poisson distribution in traffic flow, since they provide information about headways. Realistically, the hypothesis of random traffic distributions best describes situations where the traffic flow is light and vehicles can pass freely. Then the vehicles can be considered to be approximately randomly distributed along a road. In practice, however, as drivers well know, passing can be partially restricted by other vehicles in the passing lane(s), bends in the road, brows of hills, and so on. Some drivers who catch up to cars moving slightly less fast are content to stay behind them (though I find it a little frustrating), so bunching of vehicles is very common. Under these circumstances other distributions are more relevant to understanding traffic flow: for example the intervals between the "endpoints" of such bunches may follow a negative exponential or other distribution.

Chapter 10

TRAFFIC IN THE CITY

I hooked up my accelerator pedal in my car to my brake lights. I hit the gas, people behind me stop, and I'm gone.

—Steven Wright

X = q: KINEMATICS IN THE CITY

There is a fundamental relationship between the flow of traffic q in vehicles per unit time, the concentration k in vehicles per unit distance, and the speed u of the traffic. It is $q = ku$. In general each of these quantities is a function of distance (x) and time (t), but the form $q(k) = ku(k)$ may also be valuable. Another useful quantity is the spacing per vehicle, $s = k^{-1}$. If s_0 is the minimum possible spacing, that is, when the vehicles are stationary (or almost so), then $k_j = s_0^{-1}$ is referred to as the *jam concentration*; but it has nothing to do with preservatives! Associated with the above "fundamental relationship" $q = ku$ is, not surprisingly, a "fundamental diagram." The overall features can be inferred as follows. When the concentration is zero, the flow must be zero, so $q(0) = 0$. Furthermore, the flow is zero when $k = k_j$, so $q(k_j) = 0$. Since $q \geq 0$, ruling out the trivial case $q \equiv 0$ there must be an absolute maximum $q = q_{max}$ somewhere in the interval $(0, k_j)$. This is obviously of interest to traffic engineers (and indirectly, to those in traffic).

There may be more than one relative maximum of course, but the case we'll examine will have a single maximum—the "capacity" of the road. Figure 10.1 shows a typical q-k diagram for this situation.

Suppose we take two measurements of the flow $q(x, t)$ a short distance Δx apart, at points A and B, respectively (traffic moving from A to B). The flow is defined to be the number of vehicles passing a given location in time Δt; hence the change in q between the points A and B is given by

$$\Delta q = \frac{N_B}{\Delta t} - \frac{N_A}{\Delta t} \equiv \frac{\Delta N}{\Delta t}.$$

Within Δx the change in the traffic density (or concentration) k is

$$\Delta k = -\left(\frac{N_B}{\Delta x} - \frac{N_A}{\Delta x}\right) \equiv -\frac{\Delta N}{\Delta x}.$$

To see this, suppose without loss of generality that $N_A > N_B$, meaning that there is a build-up of cars between A and B, so the density in that spatial interval increases, that is, $\Delta k > 0$, as indicated above in this case. It is assumed of course that there is no creation or loss of cars from within the interval (no white holes, sinkholes, or UFO abductions)—the total number of cars is constant. From these two equations for ΔN we see that

$$\Delta q \Delta t + \Delta k \Delta x = 0,$$

or

$$\frac{\Delta q}{\Delta x} + \frac{\Delta k}{\Delta t} = 0.$$

Now we invoke the continuum hypothesis (some shortcomings of which are discussed in Chapter 15 and Appendix 8); we assume the quotients above possess well-defined limits as the discrete increments tend to zero, thus obtaining the limiting equation

$$\frac{\partial q}{\partial x} + \frac{\partial k}{\partial t} = 0. \tag{10.1}$$

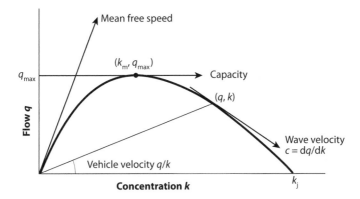

Figure 10.1. Flow-concentration diagram. The maximum is at the point of tangency (k_m, q_{max}).

This is the *equation of continuity* for the kinematic model. Note that it can be adapted to include the effects of entrances, exits and intersections, etc. by adding a term, $g(x,t)$ say, to the right hand side. We will examine some simple consequences of equation (10.1); suppose that $q = q(k)$; assuming the differentiability of q we have that

$$\frac{dq}{dk}\left(\frac{\partial k}{\partial x}\right) + \frac{\partial k}{\partial t} = 0. \qquad (10.2)$$

In the simplest possible case $dq/dk = c$, a constant, so the resulting equation is

$$\frac{\partial k}{\partial t} + c\frac{\partial k}{\partial x} = 0. \qquad (10.3)$$

This has the general solution

$$k(x,t) = h(x - ct), \qquad (10.4)$$

as is readily confirmed using the chain rule. In equation (10.4), h is a differentiable but otherwise arbitrary function. Its form depends on the so-called "initial conditions" at time $t = 0$ (say).

The solution (10.4) implies that k (and therefore q) travels to the right with "shape" h and speed c. If we recall that the mean speed of vehicles at a point is $u = q/k$ then it follows that

$$c = \frac{dq}{dk} = \frac{d(ku)}{dk} = u + k\frac{du}{dk}. \qquad (10.5)$$

This is directly analogous to the relationship between the speed of individual waves in a medium and the speed of a group of them (a "wave packet") in fluid dynamics. If u increases with traffic density (unlikely), then $c > u$, whereas if u decreases with k, the speed of the "wave" $c < u$. This gives some basic insight into traffic jams: although u is non-negative, a sufficiently negative value of du/dk can render $c < 0$; that is, changes in the traffic conditions propagate backward at speed $|c|$.

We may also infer some qualitative traffic behavior under the very reasonable assumption that the flow is in general a decreasing function of traffic density k. Now suppose that k decreases gradually in the forward direction, that is, $k'(x) < 0$. Then the front region of the flow moves out faster than the regions behind, the decrease in density is smeared out, and provided the range of k-values are less than k_m (when $q = q_{max}$), everything propagates in the forward direction. If this is not the case some of the traffic flow conditions may propagate backward.

By contrast, if there is a sharp decrease in k, such as at a traffic signal red light, the road behind the light will have a high density while that ahead will be empty of vehicles for some distance. The density behind is $k = k_j$, the jam density, while that ahead is $k = 0$. Consequently there is an abrupt change when the light changes to green. We can think of this discontinuity as being a highly compressed collection of all possible k-values, which starts to "decompress" as traffic begins to flow. Each value of the density then propagates away at its own speed; those corresponding to $q > q_{max}$ travel backward, those for which $q < q_{max}$ travel forward, and since $c = q'(k_m) = 0$ at $q = q_{max}$, the latter do not move at all. Referring to Figure 10.1 we see that the fastest forward speed corresponds to the tangent slope at $k = 0$ (a clear road ahead) and the fastest backward speed to that at $k = k_j$. This is the rear boundary along which the traffic jam resolves.

At the traffic light, $k = k_m$ corresponding to capacity flow, since $q'(k_m) = 0$. This means that a traffic signal is useful for determining the capacity flow by allowing jam conditions to build up before turning to green. If R is the duration of the red phase, G that of the green, and vehicles arrive at the light at a rate q_a and leave at a rate q_l, the total number of cars arriving in a complete cycle is $q_a(R + G)$. The condition for them all to pass through the junction during the succeeding green phase is $q_a(R + G) < q_l G$. Therefore the capacity of the junction is given by the upper bound $Gq_l/(R + G)$.

If the density k increases with forward distance there will be a "pile-up" problem, because the flow is greater for lower concentrations. This results—hydrodynamically speaking—in a *shock wave*: a rapid transition from light traffic in the rear to heavy traffic in the front. No doubt we have all experienced this. Suppose that the set $\{q_>, k_>\}$ describes the conditions just ahead of the shock front, and $\{q_<, k_<\}$ behind it, and the shock moves at speed U. The rate at which vehicles emerge from the front of the shock wave is $q_> - Uk_>$; the rate at which they enter from the rear is $q_< - Uk_<$. The number of vehicles is conserved, so these rates are equal, so

$$U = \frac{q_< - q_>}{k_< - k_>}. \tag{10.6}$$

This is the slope of the chord on the q-k diagram for a given x-value joining the shock wave entry and exit points. Note that in the limit as these differences tend to zero, the chord becomes the tangent line, and for that value of x

$$U = c = q'(x).$$

In summary, we have seen that according to the kinematic theory of traffic flow, the front of heavy traffic concentration tends to smooth out , whereas the rear steepens and forms a jam. A vehicle approaching from the rear encounters the heavy traffic suddenly, but exits it gradually. But we all know that from personal experience! And below we put a little more mathematical meat on those most annoying events, namely:

$X = d$: TRAFFIC SIGNAL DELAYS

Let's start this subsection off with a bang:

$$d = \frac{c(1-\lambda)^2}{2(1-\lambda x)} + \frac{x^2}{2q(1-x)} - 0.65\left(\frac{c}{q^2}\right)^{1/3} x^{2+5\lambda}. \tag{10.7}$$

So there! What on earth does this formula mean? It is based on a model (Webster 1958) for the average delay per vehicle (d) at an intersection controlled by a traffic signal, and unfortunately it is too complicated to derive here. It was originally formulated for traffic in the UK and has been described as one of the most influential and useful results in this field of traffic control, known now as the *Webster delay formula*, so we had better pay it some attention! The various parameters on the right-hand side of equation (10.7) are listed below:

c = cycle length, i.e., length of one complete sequence of phases (seconds);

λ = "g/c", i.e., the proportion of the cycle that is "effectively green"; *

g = "green time" (seconds)

q = flow, i.e., average number of vehicles/s (v/s);

s = saturation flow, i.e., maximum capacity of road in vehicles/s;

x = $q/\lambda s$, the degree of saturation; $0 < x < 1$; if the light were continually green, $\lambda = 1, q = s$, and therefore $x = 1$.

It should be pointed out that the light sequence in the UK is red, red and amber (i.e., orange) together, green, amber, red, etc. The "effective green time" is (green + amber − 2) seconds, the 2 seconds being an allowance for delay in starting once the signal is green. The effective green time is easily adapted to the U.S. system in which the sequence is red, green, yellow, red, etc.

In equation (10.7) the first term represents the delay to the vehicles assuming a uniform arrival rate. The second term, a correction to the first, is the additional delay due to the randomness of vehicle arrivals. It is related to the probability that sudden surges in vehicle arrivals may cause temporary "oversaturation" of the signal operation. The third term, a subtractive one, is an empirical correction factor to correct the delay estimates consistent with observational data. It amounts to about 10% of the sum of the other terms. It is not surprising therefore that the following simplification to equation (10.7) is commonly used, namely,

$$d = 0.9 \left[\frac{c(1-\lambda)^2}{2(1-\lambda x)} + \frac{x^2}{2q(1-x)} \right]. \tag{10.8}$$

Webster tabulated the delays (according to equation (10.7) for s using increments $\Delta s = 300$ in the range $[900, 3600]$; for c ($\Delta c = 5$) in the range $[30, 60]$, and ($\Delta c = 10$) in the range $[60, 120]$; for q ($\Delta q = 25$) in the range $[50, 1200]$ and for g ($\Delta g = 5$) in the range $[10, 100]$. Let us calculate d from equation (10.8) for the values

$$c = 60 \text{ s}, \lambda = g/c = 30/60 = 1/2, s = 1800 \text{ vph} = 1/2 \text{ vps},$$
$$q = 600 \text{ vph} = 1/6 \text{ vps and } x = q/\lambda s = 600/900 = 2/3.$$

Then

$$d = 0.9 \left[\frac{60\,(1/2)^2}{2\,(2/3)} + \frac{4/9}{(1/3)^2} \right] = 0.9 \left(\frac{61}{4} \right) \approx 14 \text{ s}.$$

Webster also derived two expressions for the average length L of a traffic line (or queue) at the beginning of a green phase (usually the maximum length in the cycle). The first and least accurate one is

$$L = \max \left\{ q \left(\frac{r}{2} + d \right), qr \right\},$$

where the new parameter r is the "red time." This underestimates the length by 5–10% because it is based on the assumption that vehicles do not join the line until they have reached the stop line. A more accurate formula incorporating three new parameters is

$$L = \max \left\{ q \left(\frac{r}{2} + d \right) \left(1 + \frac{qy}{av} \right), qr \left(1 + \frac{qy}{av} \right) \right\},$$

that is, each term is increased by the factor $\left(1 + \frac{qy}{av} \right)$, where y is the average spacing between vehicles in the line, a is the number of lanes and v is the "free running speed" of the traffic.

X = oh no!: TUNNEL TRAFFIC IN THE CITY

How does the fundamental diagram "square" with known traffic configurations in tunnels (where the likelihood of traffic jams is quite high, especially during rush hour)? The Lincoln Tunnel (under the Hudson River) in New York City has a maximum traffic flow of about 1600 vph (vehicles per hour) at a density of about 82 vehicles per mile moving at a mean speed of 19 mph (see Table 10.1 and the scatter plot in Figure 10.2). As noted, if the density is almost zero, the traffic flows at the maximum speed, and for a range of small densities remains nearly constant, so the flow varies almost linearly with density in this range, $q \approx ku_{max}$. This is consistent with measurements in both the Lincoln Tunnel and other NYC tunnels such as the Holland and Queens Midtown Tunnels. As the density increases, with a corresponding increase in flow, these

TABLE 10.1

Velocity (mph)	Density (cars/mile)	Traffic flow rate (cars/hr)
32	34	1088
28	44	1232
25	53	1325
23	60	1380
20	74	1480
19	82	1558
17	88	1496
16	94	1504
15	94	1410
14	96	1344
13	103	1339
12	112	1344
11	108	1188
10	129	1290
9	132	1188
8	139	1112
7	160	1120
6	165	990

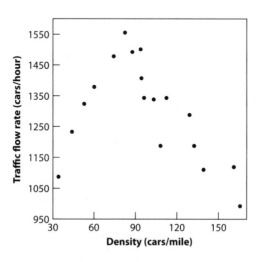

Figure 10.2. Scatter plot of data from Table 10.1.

same measurements show differences in the q-k curves: both the capacity and the corresponding speed are lower in the older tunnels (Haberman 1977). This should not be too surprising given that newer tunnels typically have greater lane width and improved lighting that permit higher speeds at the same traffic concentrations. Furthermore, the capacity of a given tunnel may vary along the road—regions with lower capacity than elsewhere are called bottlenecks, and in tunnels they typically occur on the "upward and outward" part of the tunnel (why do you think this is?).

Ideally, traffic should be forced somehow to maintain maximum flow by manipulating the density and speed appropriately. In my (limited) experience this works well on the M25 "orbital" motorway around London, which has variable speed limits (and speed cameras) during rush hour. In the Holland Tunnel, momentarily stopping traffic resulted in increased flow! This is because a traffic signal at the entrance to the tunnel was suitably timed to permit traffic to flow in intervals with density corresponding to the maximum flow.

Chapter 11

CAR FOLLOWING IN THE CITY—I

Don't some cars inevitably follow others, and not just in the city? They certainly do, but the phrase as used here means that we model traffic by identifying each car as a separate object, not just part of the flow of a fluid called "traffic." We'll start by setting up a particular type of differential equation for this (now) discrete system.

$X = x_n(t)$: A STEADY-STATE CAR-FOLLOWING MODEL

Suppose that the position of the nth car on the road is $x_n(t)$. If we disallow passing in this model we can assume that the motion of any car depends only on that of the car ahead. A simple approach is to set the car's acceleration proportional to the relative speed between it and the car in front; thus we have

$$\frac{d^2 x_n(t)}{dt^2} = -b\left(\frac{dx_n(t)}{dt} - \frac{dx_{n-1}(t)}{dt}\right), b > 0. \tag{11.1}$$

This means that if the nth car is traveling faster than the one in front, it must decelerate to avoid a collision (generally a good idea). Conversely, if it is not traveling as fast as the one in front, the driver will (in this model) accelerate accordingly. Note that the model, simplistic as it is, does not give the driver the choice of maintaining a constant speed unless the right-hand side of equation (11.1) is zero. Additionally, the equation takes no account of the time lag due to the reaction time (T) of the driver in responding to the changing conditions ahead of her.

(At this point the reader may throw up his hands in disgust and say—"These mathematicians! Nothing is ever realistic—when do such conditions *ever* occur?" He has my sympathies, but this is the way modeling is usually done: take the simplest nontrivial situation and see what the implications are for the real-world problem, and modify, tweak, and improve as necessary . . . trial and error are important in constructing models.)

Incorporating the reaction time T results in the modification

$$\frac{d^2 x_n(t+T)}{dt^2} = -b\left(\frac{dx_n(t)}{dt} - \frac{dx_{n-1}(t)}{dt}\right). \tag{11.2}$$

According to one source, T is approximately 1.5 seconds for half of all drivers, and in the range 1–2.2 sec for all drivers, though over two seconds seems rather high to me. Equation (11.2) is actually a system of *delay-differential* equations $(n = 1, 2, 3, ...)$ and in general these are notoriously difficult to solve. It can be integrated directly however, to yield

$$\frac{dx_n(t+T)}{dt} = -b\left(x_n(t) - x_{n-1}(t)\right) + c_n, \tag{11.3}$$

where c_n is a constant of integration. This equation defines the speed of the nth car in terms of the separation from the car in front at an earlier time. Let's examine the special case of a *steady state* (or time-independent) situation in which all the cars are spaced equidistantly, and hence moving at the same speed. Then, since

$$\frac{dx_n(t+T)}{dt} = \frac{dx_n(t)}{dt},$$

equation (11.3) may be reformulated as

$$\frac{dx_n(t)}{dt} = -b\big(x_n(t) - x_{n-1}(t)\big) + c_n. \qquad (11.4)$$

We may now ask how the spacing $x_n(t) - x_{n-1}(t) = -d$ might depend on the traffic concentration k. In the situation described by equation (11.4), consider all vehicles to have length l, so that the number of cars per mile (or km) will be the constant value $k = (l + d)^{-1}$. From equation (11.4) with $c_n = c$

$$u = b\Big(\frac{1}{k} - l\Big) + c.$$

Imposing the reasonable requirement that at the maximum possible density k_{max} (bumper-to-bumper traffic), $u = 0$, we can solve for the constant c to obtain the simple result

$$u(k) = b\Big(\frac{1}{k} - \frac{1}{k_{max}}\Big). \qquad (11.5)$$

There is a problem, however; this equation predicts that $u \to \infty$ as $k \to 0(!)$. But it is easily resolved, because we know that for small enough densities, $0 < k < k_c$ say, $u \approx u_{max}$. By requiring $u(k)$ to be continuous it is necessary to choose

$$k_c = \frac{bk_{max}}{b + k_{max}u_{max}}$$

so that the traffic flow is

$$q(k) = ku_{max}, \; k < k_c, \text{ and}$$
$$q(k) = b\Big(1 - \frac{k}{k_{max}}\Big), \; k \geq k_c. \qquad (11.6)$$

A typical q-k graph is shown in Figure 11.1. It is a piecewise linear approximation to the concave-down curve discussed in the previous model. One unfortunate feature is that the maximum flow occurs at $k = k_c$, which is unlikely to be the case. The corresponding u-k graph is shown as a dotted line.

The above model is rather limited in its scope, and can be improved somewhat by modifying the proportionality parameter b. This is likely to depend

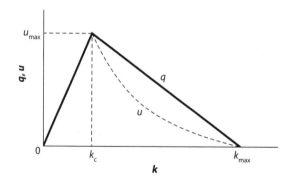

Figure 11.1. $q(k)$ and $u(k)$ profiles based on equations (11.6) and (11.5), respectively.

on the distance between the car and the one it is following; it seems reasonable to conclude that the closer it follows, the larger will be the accelerative or decelerative response. This quantity is in effect a sensitivity term. To this end, we choose

$$b = \frac{B}{x_{n-1}(t) - x_n(t)},$$

so that instead of the linear equation (11.2) we have the following nonlinear version:

$$\frac{d^2 x_n(t+T)}{dt^2} = B \frac{\left(\dfrac{dx_n(t)}{dt} - \dfrac{dx_{n-1}(t)}{dt} \right)}{x_n(t) - x_{n-1}(t)}. \tag{11.7}$$

As above, this can be integrated, yielding

$$\frac{dx_n(t+T)}{dt} = B \ln \left| x_n(t) - x_{n-1}(t) \right| + c_n. \tag{11.8}$$

Again, we choose a steady-state traffic flow for which this reduces to

$$u = B \ln \left[\frac{k_{max}}{k} (1 - kl) \right]. \tag{11.9}$$

Since

$$k = \frac{1}{l+d} < \frac{1}{l},$$

we simplify this to the case when the traffic density is much less than the bumper-to-bumper density $k_{max} = 1/l$ (for which $u = 0$), that is, $k \ll 1/l$. Then (11.9) reduces to

$$u = -B\ln\left(\frac{k}{k_{max}}\right) \tag{11.10}$$

upon which, as before, we impose the low density condition $u = u_{max}$ to avoid singular behavior as $k \to 0$. The flow is therefore given by

$$q(k) = ku_{max}, \ k < k_c, \text{ and}$$
$$q(k) = -Bk\ln\left(\frac{k}{k_{max}}\right), \ k \geq k_c. \tag{11.11}$$

For u to be continuous the choice of

$$B = \frac{u_{max}}{\ln\left(k_{max}/k_c\right)}$$

must be made, but in this model the coefficient B has a more interesting interpretation. From equation (11.10) the maximum flow occurs when

$$\frac{dq}{dk} = -B\left[\ln\left(\frac{k}{k_{max}}\right) + 1\right] = 0, \tag{11.12}$$

and this occurs when $k = k_{max}/e$. At this value of k, the speed of traffic is, from (11.10) simply $u(k_{max}/e) = B$. For consistency with the continuity requirement, k_c must be such that

$$u_{max} = u\left(\frac{k_{max}}{e}\right)\ln\left(\frac{k_{max}}{k_c}\right),$$

or more explicitly,

$$k_c = k_{max}\exp\left(-u_{max}/B\right). \tag{11.13}$$

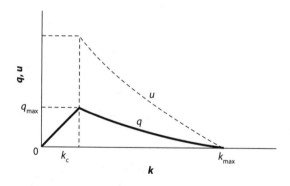

Figure 11.2. $q(k)$ and $u(k)$ profiles based on equations (11.11) and (11.10), respectively.

A generic sketch of both $u(k)$ (dotted line) and $q(k)$ is shown in Figure 11.2.

Chapter 12

CAR FOLLOWING IN THE CITY—II

I love to watch clouds; their changing forms are indicative of the different kinds of hydrodynamical process that are present in the upper atmosphere, such as convection, shear flow, and turbulence. Unfortunately, I am rather prone to do this while driving. Probably the worst example of this occurred many years ago when my wife and I were on our way to the local hospital (she was in labor with our third child). I won't elaborate here, except to say that she rightly urged me to concentrate on the road. Distractions such as cloud-watching while driving increase the reaction time for avoiding traffic hazards (and therefore should *not* be engaged in!). This next set of models incorporate reaction times in a simple and rather natural manner.

X = *Q*: ALTERNATIVE CAR-FOLLOWING MODELS

We now consider a driver traveling at speed u who tries to maintain a constant distance between his car and the one ahead of him. Since he will wish to be able to stop suddenly if the vehicle ahead does, and to do so without hitting it, the spacing s can be written (in particular) as a quadratic function of speed. (Why is this so?) Consider the expression

$$s = s_0 + au + bu^2. \tag{12.1}$$

From equation (12.1) the constants s_0, a, and b must have dimensions of distance, time, and (acceleration)$^{-1}$, respectively. Therefore s_0 might be, for instance, the minimum spacing from the back of car n to the front of the following car $(n-1)$ or front to front. The constant a could be the reaction time to a sudden braking of the car ahead, and b could be the maximum deceleration, which would modify the speed u in order to keep s constant. We can solve equation (12.1) for u:

$$u = \frac{-a \pm \left[a^2 - 4b(s_0 - s)\right]^{1/2}}{2b} = \frac{\alpha}{2\beta}\left\{-1 \pm \left[1 + \frac{4b}{a^2}(s - s_0)\right]^{1/2}\right\}, \; s \geq s_0,$$

where $\alpha = a/s_0$ and $\beta = b/s_0$. The positive solution (+ root) can be recast directly into a form related to $q = q(k)$ by defining some new parameters:

$$q = ku = k_j u\left(\frac{k}{k_j}\right) \equiv k_j uK, \; k_j = \frac{1}{s_0}, \; q_0 \equiv \frac{k_j \alpha}{2\beta}, \; \gamma \equiv \frac{4\beta}{\alpha^2} \tag{12.2}$$

from which we obtain the expression

$$q = q_0\left\{-K + \left[K^2 + \gamma K(1-K)\right]^{1/2}\right\}. \tag{12.3}$$

In view of the above discussion about the shape of the $q(k)$ (and now the $q(K)$) graph, we seek the location of the maximum from $dq/dK = 0$, that is, where

$$\left[K^2 + \gamma K(1-K)\right]^{1/2} = K(1-\gamma) + \frac{\gamma}{2}.$$

Simplifying, we obtain the following quadratic equation in K:

$$\gamma(\gamma-1)K^2 - \gamma^2 K + \frac{\gamma^2}{4} = 0. \tag{12.4}$$

This equation has the roots

$$K = \frac{\gamma \pm \sqrt{\gamma}}{2(\gamma-1)}.$$

On substituting these into equation (12.3) it transpires that only the $-$ root gives $q > 0$. We can write this value as

$$q_{max} = \frac{q_0 \gamma}{2(1+\sqrt{\gamma})} = \frac{k_j}{\alpha + 2\sqrt{\beta}}. \tag{12.5}$$

Finally, writing $Q = q/q_{max}$ we have an expression for $Q(K)$, that is,

$$Q = \frac{2(1+\sqrt{\gamma})}{\gamma}\{-K + [K^2 + \gamma K(1-K)]^{1/2}\}. \tag{12.6}$$

A typical graph of $Q(K)$ is shown in Figure 12.1. The maximum occurs at

$$K_m = \frac{\gamma - \sqrt{\gamma}}{2(\gamma-1)},$$

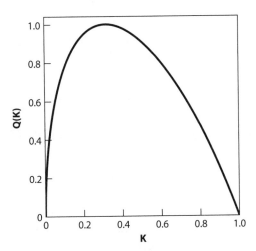

Figure 12.1. Modified q-k diagram based on equation (12.6).

The position of the maximum changes relatively little with γ; in the interval $1 < \gamma < \infty$, for example, $0.25 < K_m < 0.5$.

Note also that

$$Q'(K) = \frac{2(1+\sqrt{\gamma})}{\gamma}\left\{-1 + \frac{(1-\gamma)K + \gamma/2}{[K^2 + \gamma K(1-K)]^{1/2}}\right\}. \qquad (12.7)$$

This explains the shape of the graph near the origin; as $K \to 0^+$, $Q'(K) \to \infty$ (as does $q'(k)$ of course). As the concentration goes to zero, any interaction between vehicles will become negligible, and the flow would become an average "free speed," which renders this car-following model inapplicable. In practice, the real flow at any concentration is probably less than this, so the slope of the graph near the origin will not be so steep (see Figure 10.1).

In closing this section, let us revisit equation (12.5) for the capacity q_{max}. In terms of the original parameters (see equation (12.1))

$$q_{max} = \frac{1}{a + 2\sqrt{bs_0}}.$$

It is clear from this formula that the capacity can be raised by decreasing the reaction time (a), decreasing the headway distance (s_0), or increasing the desired maximum deceleration (b^{-1}) (not a good idea), or indeed, any combination of these changes. Conversely, the capacity will be lowered by changing them in the opposite directions.

It is appropriate to mention an "equation" from the field of psychology in this context, namely "*Response = Sensitivity × Stimulus.*" The response is usually identified as the acceleration (or deceleration) of the following vehicle, and experimental studies have shown that there is a high correlation between a driver's response and the relative speed of the vehicle ahead—the stimulus. With this in mind, note that differentiation of equation (12.1) with respect to t yields

$$\frac{du}{dt} = \left(\frac{1}{a + 2bu}\right)\frac{ds}{dt}.$$

This is a type of stimulus-response equation, although the sensitivity term $(a + 2bu)^{-1}$ decreases with increasing speed u, so it is not a particularly useful interpretation in this regard.

$X = u_n$: AN IMPROVED REACTION-TIME MODEL

There is another avenue worth pursuing in connection with the reaction time T. If we define $s_n = x_{n-1} - x_n$ as the spacing ahead of the nth car (at location x_n), rewrite equation (11.3) in terms of the discrete speed $u_n = dx_n/dt$, and let $t \to t - T$, this equation becomes (with $c_n = 0$)

$$\frac{du_n(t)}{dt} = b\frac{ds_n(t-T)}{dt}. \tag{12.8}$$

This differential-difference can be solved exactly using more advanced techniques, but we will use an approximate method consistent with the level of this book. We can expand the right-hand side of this equation as a Taylor series about t; retaining up to the quadratic term in T we obtain the expression

$$\frac{du_n(t)}{dt} \approx b\left[\frac{ds_n(t)}{dt} - T\frac{d^2s_n(t)}{dt^2} + \frac{T^2}{2}\frac{d^3s_n(t)}{dt^3}\right]. \tag{12.9}$$

Now

$$s_n = x_{n-1} - x_n, \text{ so}$$

$$\frac{ds_n}{dt} = u_{n-1} - u_n, \text{ and}$$

$$\frac{d^2s_n}{dt^2} = \frac{d}{dt}(u_{n-1} - u_n),$$

and so on. Placing all the terms involving u_n on the left-hand side, and all those in u_{n-1} on the right, we have the following constant coefficient second-order nonhomogeneous ordinary differential equation (quite a mouthful):

$$\frac{bT^2}{2}\frac{d^2u_n}{dt^2} + (1 - bT)\frac{du_n}{dt} + bu_n = b\left[\frac{T^2}{2}\frac{d^2u_{n-1}}{dt^2} - T\frac{du_{n-1}}{dt} + u_{n-1}\right]. \tag{12.10}$$

If we know the speed of vehicle $(n - 1)$, this equation enables us in principle to find the speed of the following vehicle (n). From the elementary theory of differential equations we know that the general solution of this equation is the sum of the solution to the homogeneous equation (i.e., with zero right-hand side) and a particular solution to the complete equation (with the right-hand side being a known function of t). The former solution is the most important determinant of the *instability* of the traffic flow, so we seek solutions of equation (12.10) in the form $u_n(t) = Ue^{pt}$, where U is a constant. There are two values of the constant p to be found by substitution. These values are roots of the quadratic equation

$$\left(\frac{bT^2}{2}\right)p^2 + (1 - bT)p + b = 0, \text{ i.e.,}$$

$$p_\pm = \frac{bT - 1}{bT^2} \pm \left[\frac{(bT - 1)^2}{b^2 T^4} - \frac{2}{T^2}\right]^{1/2}. \tag{12.11}$$

The nature of p (and hence the solution) depends on the sign of the radicand; it will be real (and hence non-oscillatory), provided

$$\frac{(bT - 1)^2}{b^2 T^4} - \frac{2}{T^2} > 0, \text{ or } bT < \sqrt{2} - 1 \approx 0.414.$$

Otherwise the roots are complex conjugates, and the sign of the real part determines the solution behavior. If

(i) $0 \le bT \le \sqrt{2} - 1$; both roots are negative and the solution is exponentially decreasing (stable).

(ii) $\sqrt{2} - 1 < bT < 1$; both roots are complex, and the solution is damped oscillatory (stable).

(iii) $bT = 1$; both roots are pure imaginary, and the solution is oscillatory (neutrally stable).

(iv) $bT > 1$; both roots are complex, and the solution is increasing and oscillatory (unstable).

What does all this mean? Remember that the solution $u_n(t) = Ue^{pt}$ is the speed of the nth vehicle in the line of traffic. From cases (i)–(iv) we see that (locally at least) the speed can decrease, oscillate about a decreasing mean value, oscillate about a constant mean value, and oscillate about an increasing mean value. This last case is indicative of the potential for collisions somewhere down the traffic line. The model is a crude one, to be sure, but this latter behavior is also consistent with the solution for the nonhomogeneous equation, known as the "particular integral." The choice of functional form sought depends on that of the presumed known quantity $u_{n-1}(t)$, but for our purposes it is sufficient, in light of (i)–(iv) above to consider an oscillatory solution of the form $u_n(t) = U_n \exp(i\omega t)$, $\omega \neq p_{\pm}$. Substituting this in equation (12.10) results in the expression

$$U_n\left[-\frac{b\omega^2 T^2}{2} + i\omega(1 - bT) + b\right] = bU_{n-1}\left[-\frac{\omega^2 T^2}{2} - i\omega T + 1\right]. \quad (12.12)$$

Note that

$$\left|\frac{u_n}{u_{n-1}}\right| = \left|\frac{U_n}{U_{n-1}}\right| = \left|\frac{-\frac{\omega^2 T^2}{2} - i\omega T + 1}{-\frac{\omega^2 T^2}{2} - i\omega(T - 1/b) + 1}\right|,$$

and this ratio grows as n increases if the expression on the right exceeds unity. Note that this term can be written in simplified form in terms of real and imaginary parts as

$$\left|\frac{A - iB}{A - iB(1 - 1/bT)}\right| = \left[\frac{A^2 + B^2}{A^2 + B^2(1 - 1/bT)^2}\right]^{1/2}.$$

It is clear that we require $bT > 1/2$ for $|u_n/u_{n-1}| > 1$. This then supersedes the previous criterion for instability because it "kicks in" at a lower value of bT.

For completeness in this section, the exact solution for this problem is stated below:

(i) $0 \leq bT \leq e^{-1} \approx 0.368$; both roots are negative and the solution is exponentially decreasing (stable).

(ii) $e^{-1} < bT < \pi/2 \approx 1.571$; both roots are complex, and the solution is damped oscillatory (stable).

(iii) $bT = \pi/2$; both roots are pure imaginary, and the solution is oscillatory (neutrally stable).

(iii) $bT > \pi/2$; both roots are complex, and the solution is increasing and oscillatory (unstable).

The rough and ready calculations therefore have the right "character traits" insofar as stability and instability are concerned, but the transition points are inaccurate. However, it must be said that compared with the exact solution, the upper bound in part (i) above is an overestimate of only 11%, while the transition value in part (iii) is an underestimate of about 36%.

Exercise: Show that, had we expanded equation (12.9) as a linear Taylor polynomial in T only, the results would have been very similar:

(i) $0 \leq bT \leq 1$; root is negative and the solution is exponentially decreasing (stable).

(ii) $bT > 1$; root is positive, and the solution is exponentially increasing (unstable)

for the homogeneous solution and

(iii) $bT > 1/2$ for the particular integral.

Chapter 13

CONGESTION IN THE CITY

If all the cars in the United States were placed end to end, it would probably be Labor Day Weekend.

—*Doug Larson*

$X = v(N)$: SOME EMPIRICAL MEASURES OF URBAN CONGESTION

What percentage of our (waking) time do we spend driving? In the United States, a typical drive to and from work may be at least half an hour each way, frequently more in high density metropolitan areas. So for an 8-hr working day, the drive adds at least 12–13% to that time, during much of which drivers may become extremely frustrated. (Confession: I am not one of those people; I am fortunate enough to walk to work!)

On 25 January 2011 my local newspaper carried an article entitled "Slow Motion." The article took data from the American Community Survey (conducted by the U.S. Census Bureau) and presented average commuter travel times for the region. According to the survey, the NYC metro area had the highest "average commute time," more than 38 minutes; Washington, DC, was second with 37 minutes, and Chicago was third with more than 34 minutes. (The piece in the newspaper stated the commute times to the nearest hundredth of a minute—about half a second—which is clearly silly!) The average for my locality was about 26 minutes.

In much of what follows, the equations describing various features of traffic-related phenomena are based on detailed observational studies published in the literature. Consequently, the coefficients in many of these equations are not particularly "nice," that is, not integers! For example, one measure of the capacity of a road network in a UK city center (London) was given by [25]

$$\frac{\text{Number of cars using road per hour}}{\text{Width of road in feet}} = N = 68 - 0.13v_1^2, \quad (13.1)$$

where v_1 is the average speed of traffic in mph (numerically). The traffic flow is more easily appreciated by inverting this to give the approximate expression $v_1 = (523 - 7.7N)^{1/2}$. For comparison, another empirical measure is also illustrated:

$$N = 58.2 - 0.00524v_2^3, \text{ or } v_2 = (11107 - 191N)^{1/3}.$$

As can be seen from Figure 13.1, the two formulae give similar results for speeds above about 12 mph, but in fact $v_1(N)$ (solid curve) is a better fit to the

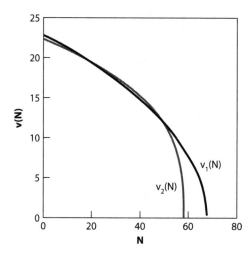

Figure 13.1. v_1 and v_2 in mph vs. N (cars/hr/highway width in ft)

low-speed data. It should be emphasized that both N and $v_{1,2}$ are numerical values associated with the units in which they are expressed, so equation (13.1) and others like it are dimensionless.

Of course, not all vehicles will be cars. In some of the literature [20]–[27], bicycles were regarded as the equivalent of one third of a car, buses the equivalent of three cars, and so forth. Thus, in order to achieve a speed of v mph, equation (13.1) suggests that each car on the road, during a period of one hour, requires a width of $(68 - 0.13v^2)^{-1}$ ft. A bus would require three times this width and a bicycle one third (a motorcycle, presumably, one half).

GENERAL ALGEBRAIC EXPRESSIONS FOR TRAFFIC DYNAMICS

$X = \delta T$: **Question:** How are increases in traffic and losses in time (due to congestion) related?

As above, we'll write $N(v)$ as a quadratic function,

$$N(v) = a - bv^2 \equiv N_0\left(1 - \frac{v^2}{v_0^2}\right), v \leq v_0. \tag{13.2}$$

where $N_0 = a$ and $v_0^2 = a/b$. N_0 is interpreted as the maximum flow of vehicles and v_0 as the speed under "light" traffic conditions. Suppose that a journey is of distance l and we define T, the accumulated time loss due to congestion, as the difference (for N vehicles) between the times traveled at speeds v and v_0, respectively, then

$$T = N(v)\left(\frac{l}{v} - \frac{l}{v_0}\right). \tag{13.3}$$

We shall define the relationship between small changes in T and N (δT and δN respectively) by

$$\delta T = T'(N)\delta N,$$

where $T'(N)$ is the derivative of T with respect to N. If T' is not too large, a small change in N induces a correspondingly small change in T according to this formula. This is a slight variation on the definition of differentials in elementary calculus, but it will be quite sufficient an approximation for our purposes, and indeed a good one, so we will use "=" rather than "≈" in what follows. From equations (13.2) and (13.3) we obtain

$$\frac{1}{l}\frac{\delta T}{\delta N} = \left(\frac{1}{v} - \frac{1}{v_0}\right) + \frac{N v_0^2}{2 N_0 v^3}, \tag{13.4}$$

or

$$\frac{\delta T}{\delta N} = \frac{l}{2v_0}\left(\frac{v_0}{v} + \frac{v_0^3}{v^3} - 2\right) = \frac{l}{2v_0}\left(\frac{v_0}{v} - 1\right)\left(\frac{v_0}{v} + \frac{v_0^2}{v^2} + 2\right). \tag{13.5}$$

As a function of the ratio v_0/v, the right-hand side of (13.5) is a cubic polynomial, increasing monotonically from zero when the traffic is light $(v_0/v = 1)$. Graphically, it is perhaps easier to appreciate as a function of the ratio of speed to the "light" speed (not the speed of light!), that is, in terms of $x = v/v_0$. Thus

$$\frac{2v_0}{l}\frac{\delta T}{\delta N} \equiv f(x) = \left(\frac{1}{x} - 1\right)\left(\frac{1}{x} + \frac{1}{x^2} + 2\right), 0 < x \le 1. \tag{13.6}$$

The graph of this decreases monotonically to zero when $x = 1$, and as can be seen in Figure 13.2, the congestion term on the left of equation (13.6) decreases by a factor of about twelve as x increases from 0.2 to 0.5; that is, when the average speed is one fifth of the speed of light traffic, the time loss is about

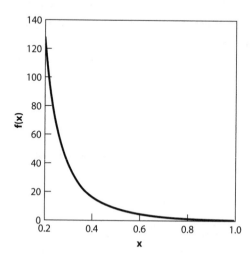

Figure 13.2. Graph of the function $f(x)$ defined by equation (13.6).

twelve times as great as when the average speed is one half that of light traffic. Furthermore, from equations (13.2), (13.3), and (13.5) we can deduce an expression for the following useful measure of congestion:

$$\frac{\text{\% increase in time loss from congestion}}{\text{\% increase in traffic}} = \frac{\delta T/T}{\delta N/N} = \frac{N}{T}\frac{\delta T}{\delta N}$$

$$= \frac{1}{2}\left(\frac{v_0}{v} + \frac{v_0^2}{v^2} + 2\right). \tag{13.7}$$

Clearly $\delta T/T$ increases as a quadratic function of the percentage increase of traffic.

$X = \delta t$: **Question:** What difference will just one more vehicle make?

The change in time for a journey of length l and duration $t = l/v$ when there is a small change δv in speed is given by $\delta t = -l\delta v/v^2$. If this change arises because of a small change δN in the traffic flow, we may write

$$\delta t = -\frac{lv'(N)}{v^2}\delta N.$$

Therefore, if N is the flow before I try to sneak my car into the traffic, the total increase in journey time of the other vehicles is approximately $N\delta t/\delta N$, that is, $-Nlv'(N)/v^2$. I do feel a little guilty about this, but I have to get to the bank before it closes. Using equation (13.2) this can be rewritten as

$$N\delta t = \frac{l}{2v}\left(\frac{v_0^2}{v^2} - 1\right). \tag{13.8}$$

We can extract more from this equation. The journey takes a time l/v, so that

$$\frac{\text{Time losses imposed on other vehicles}}{\text{Time of journey}} = \frac{1}{2}\left(\frac{v_0^2}{v^2} - 1\right). \tag{13.9}$$

As may be seen from equation (13.9), this represents a significant time loss if the speed of traffic is low. And here is another point to note: equation (13.8) represents the time loss imposed on other vehicles by my entry into the traffic stream. If we add to this the time loss sustained by my vehicle, then the resulting sum must be the same as that expressed by equation (13.5), that is,

$$\frac{l}{2v_0}\left(\frac{v_0}{v}-1\right)\left(\frac{v_0}{v}+\frac{v_0^2}{v^2}+2\right)=l\left(\frac{1}{v}-\frac{1}{v_0}\right)+\frac{l}{2v}\left(\frac{v_0^2}{v^2}-1\right). \quad (13.10)$$

Exercise: Verify this identity.

$X = dx/dt$: Question: How fast do traffic back-ups increase in length?

As we all know from experience, delays while driving can arise for several reasons: accidents or other obstructions, tunnels, bottlenecks at an intersection, raised bridges, or even a poorly adjusted traffic light at a busy junction. If the traffic flow is N vehicles at speed v, and joins a line of traffic (a traffic queue if you are reading this in the UK) with average flow N_0 at speed v_0, no line forms if $N < N_0$. By contrast, if $N > N_0$, a steadily increasing line forms (we are here excluding temporary back-ups that oscillate in length). In the latter case, the rate at which the line lengthens can be computed as follows. We denote the length of the back-up be x at time t, where $x(0) = 0$, that is, we suppose that the problem starts at time "zero." Noting that the ratio N/v has units of (vehicles/hr) ÷ (mph), or vehicles/mile, it follows that at $t = 0$, a portion of road of length x contains Nx/v vehicles. At a later time t it contains N_0x/v_0, because of the back-up. Therefore the number of vehicles entering the stretch minus the number leaving it is $(N - N_0)t$, but this is just the quantity $(N_0v_0^{-1} - Nv^{-1})x$. Hence the rate at which the back-up increases is

$$\frac{dx}{dt} \equiv \dot{x} = \frac{N - N_0}{\left(N_0v_0^{-1} - Nv^{-1}\right)}. \quad (13.11)$$

In the question below we apply this formula to an all-too-common situation in many parts of the world.

Question: Suppose the traffic is stationary. How fast is the line of traffic increasing?

This means that $N_0 = 0$ and $v_0 = 0$; clearly this means that equation (13.11) is indeterminate, so how are we to interpret the ratio N_0/v_0 in this case? Dimensionally, N/v is the ratio of vehicles per unit time to speed, that is, vehicles/length, or vehicular density on the road. In the case of stationary traffic, as here, this is just the concentration of vehicles in the queue. We shall take the *effective length* of an "average" vehicle to include the space between each one and the vehicle ahead. As before, we'll call this density k, so equation (13.11) reduces to

$$\dot{x} = \frac{N_0}{(k - Nv^{-1})}. \qquad (13.12)$$

This is the rate at which the traffic "jam" lengthens. Suppose, as an example, that we take the effective length for American vehicles (including the gap ahead) to be about 25 ft (it is probably less in Europe since the cars are typically smaller), and that the vehicles arrive at 30 mph, with a flow rate of, say, 500 vehicles per hour. The concentration per mile of the stationary line of cars, k, is then $5280 \div 25 \approx 210$ per mile. From (13.12) the line increases at a rate

$$\dot{x} = \frac{500}{210 - 500/30} \approx 2.6 \text{ mph}.$$

Obviously, one can plug in one's own estimates based on different traffic conditions.

Chapter 14

ROADS IN THE CITY

$X = \overline{L}$: **Question**: What is the average distance traveled in a city/town center?

This can be quite a complicated quantity to calculate, depending as it does on the type of road network and distribution of starting points and destinations, among other factors. Smeed (1968) assumed a uniform distribution of origins and destinations for both idealized and real UK road networks, and with some simplifying assumptions, concluded that the range of values lay between $0.70A^{1/2}$ and $1.07A^{1/2}$, with a mean of $0.87A^{1/2}$, A being the area of the town center (assuming this can be suitably defined). Obviously the factor $A^{1/2}$ renders the result dimensionally correct. With this behind us, the next stage is to try to determine the average length of a journey during say, a peak travel period. Using rather sophisticated statistical tools, Smeed was able to calculate the above average distance traveled on roads of any given town or city center, assuming only that the journeys are made by the shortest possible route. Rather than try to reproduce the (unpublished) calculations here, let us try to make these figures plausible on the basis of some simple geometric models of cities.

We will just calculate the mean lengths of parallel roads in circular and rectangular town centers as we move from one side of the town center to the other. Even this rather crude approach yields answers close to those found in the literature, indicating that the coefficient is relatively insensitive to the structure of the road network. First, we consider a set of N-S parallel roads in a circular town center of radius r (see Figure 14.1), separated by a constant distance r/n, where n is a positive integer. In so doing, we are neglecting the combined width of the road network compared with the town center area. Each road passing within a distance pr/n of the origin (where p is an integer less than n) has length $2r[1-(p/n)^2]^{1/2}$, and for each $p > 0$ there are two roads of identical length. There are $2(n-1)+1 = 2n-1$ roads, so the discrete average length is given by

$$\bar{L} = \frac{2r}{2n-1}\left[1 + 2\sum_{p=1}^{n-1}\left(1-\left(\frac{p}{n}\right)^2\right)^{1/2}\right]. \tag{14.1}$$

Of course, an identical result holds for W-E roads, and the combined average will be the same. To express $\bar{L}$ in terms of $A^{1/2}$ we merely write $\bar{L} = kA^{1/2} = k\pi^{1/2}r$, so that the coefficient $k = \bar{L}r^{-1}\pi^{-1/2}$. For $n = 3$ (five roads), $k \approx 0.99$; for $n = 4$ (seven roads), $k \approx 0.96$, and for $n = 10$ (nineteen roads), $k \approx 0.92$. Thus we see that these constants are well within the range found by Smeed.

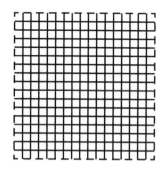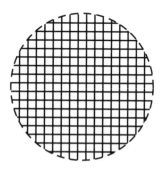

Figure 14.1. Cartesian road "grids" in rectangular (shown here, square) and circular cities.

It may be of some interest to take the mathematical limit as $n \to \infty$, in the above equation and calculate the average length for a continuum of roads—perhaps more appropriate for ducks on a circular pond (though possibly a good description of the morning commute!). In practice, it is less messy just to set up the integral for the average value of the function $L(x) = 2(r^2 - x^2)^{1/2}$, that is,

$$\bar{L} = \frac{1}{r} \int_{-r}^{r} (r^2 - x^2)^{1/2} dx = \frac{2}{r} \int_{0}^{r} (r^2 - x^2)^{1/2} dx. \qquad (14.2)$$

This is a standard integral, and can be evaluated with a trigonometric substitution to find that

$$\bar{L} = \frac{\pi r}{2}.$$

Hence, proceeding as before, setting $\pi r/2 = kA^{1/2}$ yields $k = \sqrt{\pi}/2 \approx 0.89$. It would appear that this is indeed the limit of the decreasing sequence of k-values suggested by equation (14.1).

Exercise: (a) Verify this result for $\bar{L}$; **(b)** find k for $n = 100$.

Now we proceed to evaluate $\bar{L}$ and k for a rectangular town center. We suppose that the longer sides are parallel to the x-direction, for convenience. The sides are of length a, and $b > a$. There are $n + 1$ N-S equally spaced roads (including the sides) and, similarly, $m + 1$ W-E equally spaced roads. We assume the N-S and W-E road spacing is the same, Δ, say. The average distance traveled (again, from side to opposite side, for simplicity) is

$$\bar{L} = \frac{(n+1)a + (m+1)b}{(n+1) + (m+1)}. \qquad (14.3)$$

If we let $b = (1 + \alpha)a$, this can be rearranged to give

$$\bar{L} = \left[1 + \frac{(m+1)\alpha}{m+n+2}\right]a. \qquad (14.4)$$

The parameter α is a measure of how "rectangular" the town center is; the smaller the value of α the closer the center is to being square. Furthermore, since we wish to write $\bar{L} = kA^{1/2}$ and $A = ab = (1 + \alpha)a^2$, it follows that

$$k = \left[1 + \frac{(m+1)\alpha}{m+n+2}\right](1+\alpha)^{-1/2}. \qquad (14.5)$$

This can be simplified still further by noting that, since $a = m\Delta$ and $b = n\Delta$,

$$\frac{b}{a} = 1 + \alpha = \frac{n}{m}.$$

Substituting for n we find that, after a little rearrangement,

$$k = \frac{2(1+\alpha)m+2+\alpha}{(1+\alpha)^{1/2}[(2+\alpha)m+2]}. \qquad (14.6)$$

Figure 14.2 shows excellent agreement for bounds on $k(\alpha)$ when compared with the range quoted by Smeed. In passing, it is worthwhile to note that the topics discussed here are related to the probabilistic one of determining the average distance between two random points in a circle. A simple derivation of one such result can be found in Appendix 7. There has been much in the mathematical literature devoted to this problem, and it has been adapted by several theoretical urban planning groups to model optimal traffic routes between centers of interest.

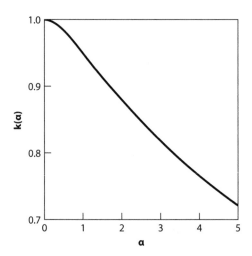

Figure 14.2. Proportionality parameter k (see equation (14.5)) as a function of α ($m = 10$ here).

$X = T_i$: **Question:** What difference does a beltway make?

A beltway (or ring road in the UK) is a highway that encircles an urban area so that traffic does not necessarily have to pass through the center. A driver wishing to get to the other side of the city without going through the center might be well advised to use this alternative route. However, I've heard it said regarding the M25 motorway (a beltway around outer London) that it can be the largest parking lot in the world at times! Nevertheless, in my (somewhat limited) experience, a combination of variable speed limits and traffic cameras usually keeps the traffic flowing quite efficiently via a feedback mechanism between the two.

As always in our simple models, circular cities will be considered to be radially symmetric; that is, properties vary only with distance r. The city is of radius r_0 and $v(r)$ is the speed of traffic (in mph) at a distance r from the city center, $0 \leq r \leq r_0$. V_0 is the speed of traffic around the (outer) beltway from any starting point A on the perimeter. For simplicity we will assume that V_0 is constant and that $v(r)$ is a linearly increasing function; as we will see, even these simplistic assumptions are sufficient to provide some insight into the potential advantages of a beltway. Thus $v(r) = ar + b$, $a \geq 0$, $b \geq 0$. A driver wishes to travel from point A to point $P(r_0, \theta)$, (see Figure 14.3) both situated on

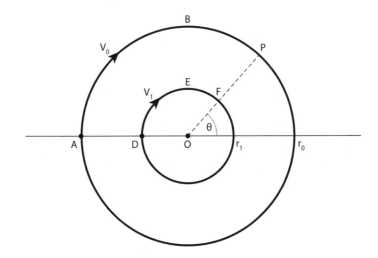

Figure 14.3. Radially symmetric velocity contours in a circular city.

the perimeter road. We shall suppose that there is also a circular "inner city" beltway a distance $r_1 < r_0$ from the center O, along which the constant speed is $V_1 \leq V_0$. She then has three choices: (1) to drive right into the city center and out again to P, the path $ADOFP$; (2) to drive around the outer beltway along path ABP, and (3) to take an intermediate route using the inner beltway along the path $ADEFP$. We shall calculate the times taken along each route under our stated assumptions.

For the first choice, noting that $v = dr/dt$, we have that

$$T_1 = \int_0^{T_1} dt = 2 \int_0^{r_0} \frac{dr}{v(r)} = 2 \int_0^{r_0} \frac{dr}{ar+b} = \frac{2}{a} \ln\left(\frac{ar_0}{b} + 1\right) \text{ hr.} \quad (14.7)$$

For the second route,

$$T_2 = \frac{(\pi - \theta)r_0}{V_0} \text{ hr.} \quad (14.8)$$

Finally, for the third route, involving some travel around the inner beltway,

$$T_3 = 2 \int_{r_1}^{r_0} \frac{dr}{v(r)} + \frac{(\pi - \theta)r_1}{V_1} = \frac{2}{a} \ln\left(\frac{ar_0 + b}{ar_1 + b}\right) + \frac{(\pi - \theta)r_1}{V_1} \text{ hr.} \quad (14.9)$$

Obviously this last case is intermediate between the other two in the sense that as the inner beltway radius $r_1 \to 0$, $T_3 \to T_1$, and as the outer radius $r_1 \to r_0$, $T_3 \to T_2$. It is interesting to compare these travel times; so let us examine the first two cases and ask when it is quicker to travel along the outer beltway to P, that is, when is $T_2 < T_1$? This inequality can be arranged as

$$f_2(r_0) = \frac{\pi a r_0}{2V_0} < \ln\left(\frac{ar_0}{b} + 1\right) = f_1(r_0). \quad (14.10)$$

Since $f_1(0) = f_2(0) = 0$, and $f_1''(r_0) < 0$, it follows that the graphs of these two functions will intersect at some radius, r_c say, if and only if $f_1'(0) > f_2'(0)$, i.e., if $2V_0 > (\pi - \theta)b$. If $r_c < r_0$ then there will be an interval $(0, r_c)$ for which it is quicker (in this model) to travel along the outer beltway to P. If $r_c > r_0$ then the curves do not intersect and it is always quicker to use the outer beltway.

Chapter 15

SEX AND THE CITY

We can think about the growth of cities in several ways, none of them prurient, despite the title of this chapter. Perhaps the most obvious one is how the population changes over time; another is how the civic area changes; yet another might be the number of businesses or companies in the city; and so on. These statistical properties are generally referred to as *demographics*, and they can include gender, race, age, population density, homeownership, and employment status, to name but a few. In this chapter we shall focus attention on some the simplest possible models of population growth in several different contexts, ending with some using "real" data from a bygone era. We'll start with exponential growth.

X = P(t): MATHEMATICAL MODELS
OF POPULATION GROWTH

In the November 4, 1960, issue of the journal *Science*, a paper [28] was published with the following provocative title: "Doomsday: Friday 13 November, A.D. 2026." Beneath this was the sentence: *At this date human population will approach infinity if it grows as it has grown in the last two millennia.*

The authors presented a formula for the population *P* in terms of positive constants *A*, *b*, and t_D (to be defined below) as

$$P(t) = \frac{A}{(t_D - t)^b}.$$

In this equation, the subscript *D* stands for "Doomsday"! We proceed to derive more simply a form of this result to illustrate the principle behind it. But first, a caveat. In what follows, differential calculus is used. So what's the problem with that? Well, in so doing, we are making the implicit assumption of differentiability, and hence continuity of the population of interest, whether it be humans, bedbugs, or tumor cells. But the populations are discrete! In these models, there are *always* an integral number of people, bedbugs, and so on. Calculus is strictly valid only when there is a continuum of values of the variables concerned and the dependent variables are differentiable functions. In that sense continuum models can never be totally realistic, even when there are billions of individuals (such as the number of cells in a tumor). Nevertheless, if there are sufficiently many individuals, we can justifiably approximate the properties of the population and its growth using calculus. Further informal discussion of one aspect of this "discrete/continuum" problem can be found in Appendix 8. Now we can begin!

Everyone has heard of "exponential growth" but not everyone knows what it means. It applies to (here, differentiable) functions *P(t)* for which the growth rate is proportional to *P*, or equivalently, the "per capita" growth rate is a constant, *k*. In mathematical terms

$$\frac{dP}{dt} = kP, \text{ or } \frac{1}{P}\frac{dP}{dt} = k. \tag{15.1}$$

If we know that at some time *t* = 0 say, the population is P_0, then the solution of equation (15.1) is

$$P(t) = P_0 e^{kt}. \tag{15.2}$$

If $k < 0$, then P decreases exponentially from its initial value of P_0. Such exponentially decreasing solutions are used in many algebra books to illustrate the phenomenon of radioactive decay. If $k > 0$ we see an immediate problem: the solution grows without bound as $t \to \infty$. This is reasonably supposed to be unrealistic, because if P represents the population of a city, country, or the world, there is not an unlimited supply of resources to maintain that growth. In fact, the English clergyman Thomas Malthus (1766–1834) wrote an essay in 1798 stating that "The power of population is indefinitely greater than the power in the earth to produce subsistence for man." This has been paraphrased to say that populations grow geometrically while resources grow arithmetically. Not surprisingly, exponential growth is sometimes referred to as "Malthusian growth."

Let's modify equation (15.1) just a little, and examine the innocuous looking first-order ordinary differential equation

$$\frac{dP}{dt} = kP^{1+\varepsilon}, \tag{15.3}$$

(with $P(0) = P_0$ as before). For reasons discussed below we will restrict the parameter ε to the range $-1 < \varepsilon < \infty$. This simple looking *nonlinear* ordinary differential equation has some surprises in store for us, and in fact is sometimes referred to as the "Doomsday" equation. The first thing to do is integrate it to obtain

$$-\frac{P^{-\varepsilon}}{\varepsilon} = kt + C, \tag{15.4}$$

where C is a constant of integration. This can be found immediately by setting $t = 0$ in equation (15.4). After a little rearrangement, the solution to equation (15.3) is found to be

$$P(t) = \frac{P_0}{\left(1 - \varepsilon P_0^\varepsilon kt\right)^{1/\varepsilon}} \equiv \frac{A}{\left(t_D - t\right)^b}, \tag{15.5}$$

with the restriction on t being

$$t < t_D = \left(\varepsilon P_0^\varepsilon k\right)^{-1}, \tag{15.6}$$

Equation (15.5) is in precisely the form stated at the beginning of this subsection.

Exercise: Derive the solution (15.5).

Before we examine some of the implications of this solution, and restrictions on it (remember: every equation tells a story), let's reexamine equation (15.3) and plot the quantity $k^{-1}dP/dt = P^{1+\varepsilon} \equiv r_i$, $i = 0, 1, 2$, as a function of P for the three cases $\varepsilon > 0$, $\varepsilon = 0$, and $-1 < \varepsilon < 0$, respectively. This last inequality ensures that the population growth rate for this case is not declining, since

$$\frac{d^2P}{dt^2} = (1+\varepsilon)kP^\varepsilon \frac{dP}{dt} = (1+\varepsilon)k^2 P^{\varepsilon(1+\varepsilon)} > 0.$$

In Figure 15.1 these values correspond to the top, middle, and bottom curves. The case $\varepsilon = 0$ (dashed line) clearly corresponds to the equation for exponential growth we have already seen (equation (15.2)). The upper curve corresponds to the arbitrary (and for illustrative purposes, rather large) value for ε of 0.5.

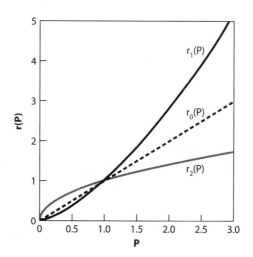

Figure 15.1. Curves proportional to the growth rates for super-exponential, exponential, and sub-exponential growth (from equation (15.3)).

The lower curve has $\varepsilon = -0.5$. Now let's consider a much smaller but positive value of ε, 0.05 say. Then the growth is just slightly "super-exponential" with the right-hand side of equation (15.3) increasing in proportion to $P^{1.05}$. From equation (15.5) the solution is

$$P(t) = P_0 \left(1 - 0.05 P_0^{0.05} kt\right)^{-20}.$$

This solution is undefined ("blows up"!) at time

$$t = t_D = \left(0.05 P_0^{0.05} k\right)^{-1} = \frac{20}{P_0^{0.05} k}, \qquad (15.7)$$

in whatever units of time are being used (usually years, of course). This doomsday time is the *finite* time for the population to become unbounded. Thus if we take the current "global village" world population of 7 billion (as declared on October 31, 2011) to be P_0, and, for simplicity, an annual growth rate $k = 0.01$ yr^{-1} (just 1% per year) then for the (admittedly arbitrary) value $\varepsilon = 0.15$, equation (15.7) gives

$$t_D = \frac{1}{0.15 \times 10^{-2} \times \left(7 \times 10^9\right)^{0.15}} \approx 22 \text{ years}.$$

This is the doomsday time for our chosen value of ε! Note that if $\varepsilon \leq 0$ the problem does not arise because there is no singularity in the solution. The population will still tend to infinity, but will not become infinite in a finite time, as is predicted for *any* $\varepsilon > 0$, *no matter how small*. The original "doomsday" paper [28] (and others listed in the references) should be consulted for the specific demographic data used.

Despite such a projection, I thought there was no suggestion that the world population is actually growing according to equation (15.3) until I read a paper [29] published in 2001. The first two sentences from the abstract of that paper may serve to whet the reader's appetite for further discussion later in this book (Chapter 18): "Contrary to common belief, both the Earth's human population and its economic output have grown faster than exponential, that is, in a super-Malthusian mode, for most of known history. These growth rates are compatible with a spontaneous singularity occurring at the same critical time 2052±10 signaling an abrupt transition to a new regime." Doomsday again?

Exercise: Obviously the choice $\varepsilon = 0$ in equation (15.3) results in the standard exponential growth of P as noted above. Using a limiting procedure in equation (15.5), show that

$$\lim_{\varepsilon \to 0} P(t) = P_0 e^{kt}.$$

Exercise: Another model, the "double exponential" model of bounded population growth, is defined by

$$P(t) = P_\infty a^{b^t},$$

where P_∞ is the asymptotic population (approached as $t \to \infty$), $0 < b < 1$, and $0 < a < 1$.

(i) Show that

$$\log\log\left(\frac{P_\infty}{P(t)}\right) = \log\log\left(\frac{1}{a}\right) + t\log b.$$

(ii) A town population starts at $P(0) = 80{,}000$ and has an upper bound $P_\infty = 200{,}000$. It is known that after 10 years the population has risen to 150,000. Determine the constants a and b.

(iii) Find the population after 15 years.

$X = x(t)$: ANOTHER SIMPLE NONLINEAR MODEL—THE LOGISTIC EQUATION

Now we shall discuss a population model that predicts a definite limit on growth. Let's examine the logistic differential equation

$$\frac{dx}{dt} = kx(N - x); \tag{15.8}$$

both k and N are positive constants. We have changed the dependent variable from P to x to reflect the fact that these models can apply to many different things. N is often called the carrying capacity in ecological contexts and the saturation level in chemistry. It can be used to describe, in a very simple manner, the spread of diseases or rumors in a closed community of population N, or a cultural fad, or indeed the spread of anything that can be spread by casual person-to-person contact, from the common cold to word-of mouth advertising about a new brand of coffee! (See Appendix 8.) The variable $x(t)$ describes the number of individuals with the disease, or who have heard the rumor, read the ad, and so on, and for our purposes the most relevant range of values for x is $0 < x < N$. The case $x > N$ is also easily accommodated and will be discussed below. Equation (15.8) is separable, and using partial fractions we can integrate it to obtain

$$\int \frac{dx}{x(N-x)} = \int \left(\frac{1}{x} + \frac{1}{N-x} \right) dx = Nkt + D,$$

D being a constant of integration to be determined. Hence

$$\frac{x}{N-x} = Ke^{Nkt},$$

where we have defined $K = e^D$. The general solution $x(t)$ is therefore

$$x(t) = \frac{NKe^{Nkt}}{1 + Ke^{Nkt}}.$$

Now rumors don't start and diseases don't spread without someone to initiate them, so we define the initial value of x to be $x(0) = x_0 > 0$. This enables us to find K in terms of the (in principle known) quantities x_0 and N. Thus

$$K = \frac{x_0}{N - x_0}.$$

On rearranging the final result, the solution becomes

$$x(t) = \frac{N}{1 + (Nx_0^{-1} - 1)e^{-Nkt}}. \tag{15.9}$$

Check that the initial condition is satisfied, and note also that $\lim_{t \to \infty} x(t) = N$. This means that $x = N$ is a horizontal asymptote; if nothing else changes, then the population x approaches the limiting value N from below (in this case) as time increases. Note that the solution (15.9) is still valid if $x_0 > N$; now the solution decreases monotonically from its initial value toward the asymptote from above. This case is meaningless in the present context, though it can apply to a simple model of overpopulation when the natural resources in a nature reserve, say, become compromised, perhaps by pollution or an environmental disaster. Then the normally sustainable population can no longer be supported, and the numbers decrease toward the *new* carrying capacity.

Further information about the evolution of $x(t)$ may readily be found from examining both its population growth rate and its *per capita* growth rate using equation (15.8). The latter rate is just

$$\frac{1}{x}\frac{dx}{dt} = k(N - x),$$

and clearly in the interval $0 < x < N$ this is a linearly decreasing function. This means that the per capita growth rate slows uniformly from a maximum near kN (when x is small) toward zero as the carrying capacity (or saturation level) is approached. On the other hand, it can be seen from equation (15.8) that the population growth rate dx/dt as a function of x is a downward facing parabola with intercepts at $x = 0, N$ and a maximum of $kN^2/4$ at $x = N/2$.

$X = ugh!$: BED BUGS (OR RATS) IN THE CITY

We can now change the context somewhat from the spread of rumors and diseases to a city-related version of harvesting or fishing. Suppose that a city has a problem with vermin—whether of the four-legged, six-legged (bed bugs!), or winged variety (e.g., pigeons), and a program is introduced to reduce the population of these unwanted "critters." How might we incorporate this "vermin reduction program" into the equations we have been discussing above? We can do this mathematically in two straightforward ways, depending on how the program is administered: by including a constant reduction rate or including a term proportional to the *existing* vermin population. Were we to replace "vermin" by "fish" then we would note that the former more appropriately describes a fishing effort in which the rate at which fish are caught on each fishing foray is the same. In the latter case the catch rate is proportional to the population, and the number of fish caught will thus diminish or increase as the population does. Taking the latter case first because it is mathematically simpler, and possibly more relevant to a town or city with a vermin problem, we can modify equation (15.8) as

$$\frac{dx}{dt} = kx(N - x) - cx, \, c > 0. \tag{15.10}$$

Here the term cx represents the vermin reduction rate (in units of e.g. rats/week). The constant c can be thought of as the per capita reduction rate (rat?) resulting from the program. Noting that equation (15.10) may be rewritten as

$$\frac{dx}{dt} = kx(\tilde{N} - x), \tag{15.11}$$

where $\tilde{N} = N - c/k$, we see that this has been encountered above; equation (15.11) is just (15.8) with a reduced carrying capacity $\tilde{N}$ (we assume that $c \le Nk$). The solution (15.9) still applies with this new horizontal asymptote replacing the earlier one. Nevertheless, this reduction is obviously helpful in the relentless fight against foul vermin! In terms of the original capacity N, the maximum growth rate is $k(N - c/k)^2/4$ and occurs at $x = (N - c/k)/2$.

Returning to the constant reduction rate, we now have a governing equation

$$\frac{dx}{dt} = kx(N - x) - c, \, c > 0. \tag{15.12}$$

To study this, we write the quadratic polynomial on the right-hand side of (15.12) as $P_c(x)$ in the form

$$P_c(x) = -k\left[\left(x - \frac{N}{2}\right)^2 - \left(\frac{N^2}{4} - \frac{c}{k}\right)\right]. \tag{15.13}$$

Now $P_c(x)$ possesses real zeros if and only if $N^2/4 > c/k$, that is, if $c < kN^2/4$. Then these zeros occur at

$$x = \frac{N}{2}\left(1 \pm \sqrt{1 - \frac{4c}{kN^2}}\right) \equiv x_\pm. \tag{15.14}$$

Clearly, $P_c(x) > 0$ when $x_- < x < x_+$ as opposed to the case for $c = 0$, that is, where $P_0(x) > 0$ when $0 < x < N$. Another way of thinking about this is by noticing that the right-hand side of equation (15.12) is just the original quadratic function $kx(N - x)$ "lowered" by the amount c. This is illustrated in Figure 15.2 for x in units of N (or equivalently, $N = 1$) and $k = 1$, $c = 0.15$.

Recalling that $P_c(x) = dx/dt$, we can easily see from the figure that for $c \ne 0$ (dotted curve), $dx/dt < 0$ for both $x < x_-$ and $x > x_+$. This means, according to

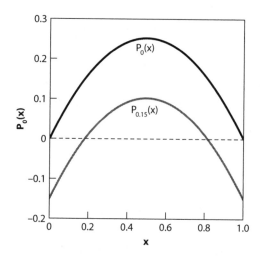

Figure 15.2. Right-hand side of equation (15.12) for $c = 0$ and 0.15 (solid and dotted curves respectively).

this simple model, that the population of rats, bedbugs, or whatever it is declines to zero or to x_+. Obviously the former is the more desirable of the two situations. Furthermore, if $c > kN^2/4$ then dx/dt for all x-values and the vermin population declines continuously to zero. This might reasonably be termed "overkill"!

Exercise: Show that the solution of equation (15.12) is given by

$$x(t) = \frac{x_+ + Qx_- e^{-dkt}}{1 + Qe^{-dkt}}, \text{ where } Q = \frac{x_+ - x_0}{x_0 - x_-} \qquad (15.15)$$

and d is the difference of the zeros of $P_c(x)$, namely, $x_+ - x_-$. When $c = 0$, $x_+ = N$, and $x_- = 0$ also verify that result this reduces as it should to the solution (15.9). The graphs of $x(t)$ in units of N are shown in Figure 15.3 for both $c = 0$ (solid curve) and $c = 0.15$ (dashed curve), with $k = 1$ and $x_0 = 0.25$. Each exhibits the classic "logistic curve" sigmoid (or stretched S) shape.

$X = \Sigma N$: HOW MANY PEOPLE HAVE LIVED IN LONDON?

Well, I was one of them, but enough about me. Suppose that at time t_1 the population was N_1, and at a later time t_2 the population was N_2. Assuming that

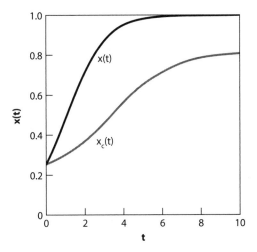

Figure 15.3. Solution curves from equation (15.15) for $c = 0$ and 0.15.

the growth was exponential, the annual growth rate r can then be determined from these data using the equation

$$N_2 = N_1 e^{r(t_2 - t_1)}, \text{ i.e.,}$$

$$r = \frac{\ln(N_2/N_1)}{t_2 - t_1}. \tag{15.16}$$

At any time t during the interval $[t_1, t_2]$ the population was $N_1 e^{r(t - t_1)}$ and the total number of "person-years" from t_1 to t_2 is therefore

$$\int_{t_1}^{t_2} N_1 e^{r(t - t_1)} dt = \frac{N_1}{r} \left[e^{r(t_2 - t_1)} - 1 \right]. \tag{15.17}$$

Hence the number of person-years lived is, from (15.16),

$$\frac{N_2 - N_1}{r} = \frac{(N_2 - N_1)(t_2 - t_1)}{\ln N_2 - \ln N_1}. \tag{15.18}$$

The left-hand side, expressed in person-years, can be converted to persons by dividing by a time factor. This should be an average life expectancy, but that is very difficult to determine, even if known in principle, owing to gender, regional, and temporal variations. For example, extracting possibly relevant (i.e., Northern European) data from a table in the *Encyclopedia Britannica* (1961), we have:

Humans by Era	Average Lifespan at Birth (years)
Upper Paleolithic	33
Neolithic	20
Bronze Age and Iron Age	35+
Medieval Britain	30
Early Modern Britain	40+
Early Twentieth Century	30–45

In view of these "data" we shall take an average of $r^{-1} = 40$ years for the range 1801–2006 considered in Table 15.1, and estimate the quantity

$$\Delta N = \frac{(N_2 - N_1)(t_2 - t_1)}{40(\ln N_2 - \ln N_1)} \qquad (15.19)$$

TABLE 15.1

Year	Population N (millions)	$\ln N$	$40\Delta N$ (units of 10^7)	ΔN (units of 10^6)
1801	0.959	13.775		
1831	1.655	14.320	3.824	0.956
1851	2.363	14.675	3.989	0.997
1891	5.572	15.533	14.960	3.740
1901	6.507	15.688	6.032	1.508
1911	7.160	15.784	6.802	1.701
1921	7.387	15.815	7.323	1.830
1931	8.110	15.909	7.691	1.921
1939	8.615	15.969	6.733	1.683
1951	8.197	15.919	10.032	2.508
1961	7.993	15.894	8.160	2.040
1971	7.453	15.824	7.714	1.929
1981	6.805	15.733	7.121	1.780
1991	6.829	15.737	6.000	1.500
2001	7.322	15.806	7.145	1.786
2006	7.657	15.851	3.722	0.931
Total				**26.810**

for a range of data points. Summing all these from the table we have the estimate of

$$N_{tot} = \sum \Delta N \approx 26.8 \text{ million people.}$$

Clearly this is a rough-and-ready approach, dependent on the width of the time intervals and the reliability of the data. The result is not particularly sensitive to the choice of life expectancy, but it *is* sensitive to the number of intervals chosen in the overall time frame. The more intervals we choose, assuming the data are available, the more different the cumulative population will generally be. To see this, just take a *single interval* from 1801 to 2006. We take $t_1 = 1801$ and $t_2 = 2006$, and $N_1 = 9.59 \times 10^5$, $N_2 = 7.66 \times 10^6$. Using equation (15.19), we find that N_{tot} (now equal to ΔN of course) ≈ 16.7 million. If we were to go back to the years $t_1 = 1000$ and $t_2 = 2600$ as before, then $N_1 \approx 10^4$ and N_2 is again as before. The corresponding value of N_{tot} is now 290 million for the same value of r. Obviously the value for London's population in the year 1000 is somewhat suspect, but this is just to illustrate the point regarding the interval width. The average lifetime was probably lower in medieval times, so if we set $r^{-1} = 30$ yr then N_{tot} increases to 386 million.

Nevertheless, this is an interesting application of some simple mathematics, and is probably far more reliable than estimates of all the people who have ever lived (Keyfitz 1976)!

Chapter 16

GROWTH AND THE CITY

Patterns of the distribution of populations around city centres are extremely variable. Clearly city development depends on constraints imposed by features of the local geography such as lakes, coastlines and mountain ranges. Rivers have played an important role as centres of attraction and indeed one of the first known cities, Babylon, was located at the junction of the Tigris and Euphrates rivers. The sizes of population centres also exhibit great variability, with numbers from just a few to ten or more million.

—A.J. Bracken and H.C. Tuckwell [30]

$X = N_{tot}$ and $X = \rho(r)$: SIMPLE URBAN GROWTH MODELS

In light of the above comments, it is somewhat surprising that fairly general quantitative patterns of urban population density ρ (population per unit area) can nevertheless be formulated for single-center cities. As far back as 1951 Colin Clark, a statistician, compiled such data for 20 cities, and found that ρ declined approximately exponentially with distance from city centers (though "city center" is not always easily defined in practice; *central business district* is perhaps a better term). If we restrict ourselves to the special but important case of circular symmetry, wherein $\rho = \rho(r)$, r being the radial coordinate, then naturally we expect that $\rho'(r) < 0$, but also perhaps that the density drops off more gradually as r increases, that is, $\rho''(r) > 0$. Furthermore, we might anticipate that this density profile gets flatter ($\rho''(r) < 0$) as the "boundary" of the city is approached. If we consider the simplest possible model of exponential decline, the suggested conditions for ρ are satisfied, so let us consider

$$\rho(r) = Ae^{-br}, r \geq 0, \tag{16.1}$$

A and b being positive constants; in fact from (16.1), $A = \rho(0) \equiv \rho_0$ and $b = -\rho'(0)/\rho_0$. Of course, in practice, urban population growth is a dynamic process, so to be more general we should permit A and b to be time-dependent (and this we shall do shortly). In fact, several values of A and b reported for London over a period of 150 years are listed below, along with correspondingly less information for three other cities, Paris, Chicago, and New York. Before analyzing these data, we derive an expression for the population.

The total metropolitan population in a disk of radius R_0 is given by the integral

$$N(R_0) = 2\pi \int_0^{R_0} r\rho(r)dr = 2\pi A \int_0^{R_0} re^{-br}dr = \frac{2\pi A}{b^2}\left[1 - e^{-bR_0}(bR_0 + 1)\right]. \tag{16.2}$$

(This is called the "civic mass" in Chapter 17.) An estimate of the total population in a spatially infinite city(!) can be found from the limiting case

$$N_{tot} = \lim_{R_0 \to \infty} N(R_0) = \frac{2\pi A}{b^2}. \tag{16.3}$$

This is not such an unreasonable result as might be first thought, because the population density (16.1) is decreasing exponentially fast. Obviously the result is formally unchanged if A and b are time-dependent; for example, if the data justified it we could choose both parameters to be linearly decreasing functions of time. This will no doubt vary from city to city and we do not pursue it here.

In Tables 16.1–16.4, A is expressed per square mile in thousands, and b in $(\text{miles})^{-1}$. Cities have two fundamental modes of growth: up and out! Of course, in practical terms growth is usually a combination: there are ultimate limits on both of them. The parameter A is a measure of "up" in the sense of central density; b is a measure of how far "out" growth occurs—the smaller b is, the more decentralized the city becomes—this is often referred to as "urban sprawl." Notice that the total population is proportional to A but inversely proportional to the square of the parameter b. The expression for N_{tot} based on equation (16.3) is generally of the right order of magnitude, particularly for London, but it does overestimate populations in later years, particularly for Chicago and New York. This partly because of the uncertainty of densities at small values of R, and also because cities are decidedly not circularly symmetric (as the first paragraph in this section implies)! Political and not just natural boundaries also play a role in urban development.

Generally, the value of the central density $\rho(0) = A$ decreased, and so did b, meaning that the population of London became more decentralized. This trend is not always realized (at least, according to the limited data listed in Montroll and Badger (1974)). The corresponding data for Chicago, Paris, and New York, are shown in Tables 16.2–16.4.

Notice the precipitous drop in both parameters for London and Rome following the end of World War II.

TABLE 16.1.
Estimated population of London in the period 1801–1951 (based on equation (16.3))

Year	A	b	N_{tot} (millions)
1801	269	1.26	1.06
1841	279	0.94	1.99
1871	224	0.61	3.78
1901	170	0.37	7.80
1921	115	0.27	9.91
1931	123	0.28	9.86
1939	83	0.22	10.77
1951	62	0.20	9.74

TABLE 16.2.
Estimated population of Chicago in the period 1880–1950 (based on equation (16.3)

Year	A	b	N_{tot} (millions)
1880	76	0.60	1.33
1900	110	0.45	3.41
1940	120	0.30	8.38
1950	68	0.18	13.19

TABLE 16.3.
Estimated population of Paris in the period 1817–1946 (based on equation (16.3))

Year	A	b	N_{tot} (millions)
1817	450	2.35	0.51
1856	240	0.95	1.67
1896	370	0.80	3.63
1931	470	0.75	5.25
1946	180	0.34	9.78

TABLE 16.4.
Estimated population of New York in the period 1900–1950 (based on equation (16.3))

Year	A	b	N_{tot} (millions)
1900	178	0.32	10.92
1910	59	0.21	8.41
1925	81	0.21	11.54
1940	110	0.21	15.67
1950	240	0.17	52.18

According to Montroll and Badger (1974), the British archaeologist Sir Leonard Woolley estimated that the ancient city of Ur had an average density of 125,000 people per square mile at the height of its mature phase, around 2000 B.C. The density per square mile for parts of fourteenth-century Paris was 140,000, as was true for parts of London in 1700. By 1900, parts of New York's Lower East Side had reached densities of 350,000 per square mile, but even this is small compared with many non-Western cites. If A increases while b remains constant, this can be accomplished quite easily over time. Thus parts of Hong Kong reached densities of about 800,000 per square mile; such high densities correspond to about one person for every four square yards!

Exercise: Verify this last statement.

One problem with the density profile (16.1) is that it cannot reproduce the so-called "density crater" for the resident population in a large metropolitan area. This phenomenon means that the maximum population density occurs, not in the central region but in a ring surrounding the city center. This can be accomplished with the function

$$\rho(r) = Ae^{br-cr^2} = \rho_0 e^{br-cr^2}, \ r \geq 0, \tag{16.4}$$

where $b > 0$, $c > 0$, and the maximum density occurs at radius $r = b/2c$. (Note that b is now of opposite sign to its counterpart in (16.1).) A graph of the normalized density function $\rho_n(r) = \rho(r)/\rho_0$ is shown in Figure 16.1 for the simple choice of $b = c = 1$.

Equation (16.4) could represent the density of a city with an extensive central business district. Let's investigate the properties of this profile in some detail. The maximum density is

$$\rho\left(\frac{b}{2c}\right) = Ae^{b^2/4c}. \tag{16.5}$$

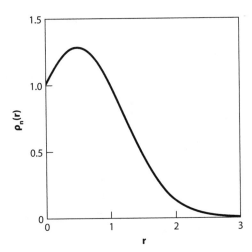

Figure 16.1. Form of the normalized population density function (16.4).

We suppose that there is a fairly well defined "perimeter" of the urban area at $r = r_p$, for which $\rho(r_p) = \rho_p$. Solving equation (16.4) for r_p yields the result (since $\rho_0 > \rho_p$ for all realistic models of cities):

$$r_p = \left[b + (b^2 + 4c\ln(\rho_0/\rho_p))^{1/2}\right]/2c. \qquad (16.6)$$

It is possible to classify these urban density profiles in terms of the magnitude and sign of the parameter combination $\beta = b/\sqrt{2c}$. Specifically, the sequence "youth, early maturity, late maturity, old age" is characterized by Newling (1969) as corresponding to the β-intervals* $(-\infty, -1)$, $(-1, 0)$, $(0, 1)$, and $(1, \infty)$. This can be seen in a qualitative manner from Figure 16.1 by mentally imagining these four stages of development to be, successively, the curve to the right of the ordinates $r = 2, 1, 0.3$, and 0. But from where do these intervals for β come? It's all to do with points of inflection. Simply put, points of inflection in the graph of $\rho(r)$ occur when there is a change of concavity (if those points are in the domain of the profile). From equation (16.4) such points occur at

$$r = r_i = \frac{b \pm \sqrt{2c}}{2c}.$$

It can be seen that if $b < -\sqrt{2c}$, i.e., $\beta < -1$, then there are no points of inflection and $\rho(r)$ decreases monotonically from the central business district outward (youthful city). If $-\sqrt{2c} < b < 0$, $(-1 < \beta < 0)$, or $0 < b < \sqrt{2c}$ $(0 < \beta < 1)$ there is a single point of inflection (in early and late maturity, respectively). Finally, if $b > \sqrt{2c}$ $(\beta > 1)$ there are two such points and the full density crater profile is evident (aging city).

There can be variations on this theme using the concept of a "traveling wave of metropolitan expansion." We can illustrate the stages of city development by positing a traveling pulse of "shape" $\rho(r)$ moving outward with speed v, that is,

$$\rho(r,t) = \rho_0 e^{b(r-vt)-c(r-vt)^2} \equiv \rho_0(\xi),\ \xi = r - ct. \qquad (16.7)$$

This can be interpreted as a point of constant density moving out radially with speed v. In principle v may itself be a function of time; in any case, as the city develops in time, the shape of the density profile may change in accordance with the above equation.

---

*where now b may be of either sign.

Returning to the basic form (16.4), we now examine the density profile when some or all of the parameters $A(=\rho_0)$, b, and c may be functions of time. In view of the *spatial* behavior exhibited by $\rho(r)$ from (16.4), that is, initially increasing to a maximum, then decreasing monotonically, Newling (1969) chose the same *temporal* form for $\rho_0(t)$:

$$\rho_0(t) = \rho_0 e^{mt-nt^2}, \tag{16.8}$$

ρ_0 now being a constant. Furthermore, he chose b to be a linear function, $b(t)$ $= b_0 + gt$, where g and c are constants. From equation (16.4) (noting again the change in sign for b from equation (16.1)) it can be seen that $b = \rho'(0)/\rho(0)$, the prime referring to a spatial derivative. Thus

$$\rho(r,t) = \rho_0 e^{mt-nt^2} e^{(b_0+gt)r-cr^2}. \tag{16.9}$$

Proceeding as with the derivation of equation (16.6) gives the corresponding result for the radius of the urbanized area (now at time t) as

$$r_p(t) = \left[(b_0 + gt) + \left((b_0 + gt)^2 + 4c \left[\ln(\rho_0/\rho_p(0)) + mt - nt^2 \right] \right)^{1/2} \right]/2c. \tag{16.10}$$

The "speed" of urban expansion can be then defined as $r_p(t)/t$ for $t > 0$. This type of model has also been used in ecological contexts, studying the spread of animal and insect populations (as well as diseases) in so called reaction-diffusion models.

Now let's use the density profile (16.4) to compute some overall population levels. As in equation (16.2), we may compute the total metropolitan population in a disk of radius R_0. It is given by the integral

$$N(R_0) = 2\pi \int_0^{R_0} r\rho(r)\,dr = 2\pi A \int_0^{R_0} re^{br-cr^2}\,dr = 2\pi Ae^{b^2/4c} \int_0^{R_0} re^{-c(r-b/2c)^2}\,dr. \tag{16.11}$$

Changing the variable of integration to $R = r - b/2c$ enables us to write $N(R_0)$ as the sum of two integrals, namely,

$$N(R_0) = 2\pi Ae^{b^2/4c} \left[I_1 + \frac{b}{2c}I_2 \right]\Big|_\alpha^\beta,$$

where the limits are now $\alpha = -b/2c$ and $\beta = R_0 - b/2c$, and the (indefinite) integrals are

$$I_1 = \int Re^{-cR^2} dR \text{ and } I_2 = \int e^{-cR^2} dR.$$

Noting that the second of the (definite) integrals can be expressed in terms of the error function

$$\mathrm{erf}(x) = \frac{2}{\sqrt{\pi}} \int_0^x e^{-r^2} dr,$$

and the expression for the total population within a disk of radius R_0 can be written as

$$N(R_0) = \frac{\pi A}{c} \left[1 - e^{R_0(b - cR_0)} + \frac{b\sqrt{\pi}}{2\sqrt{c}} e^{b^2/4c} \left\{ \mathrm{erf}(\alpha\sqrt{c}) - \mathrm{erf}(\beta\sqrt{c}) \right\} \right]. \quad (16.12)$$

Noting that $\mathrm{erf}(\infty) = 1$, we may take the formal limit of this equation to find that in the limiting case of an infinitely large city with this density profile,

$$N_{tot} = \frac{\pi A}{c} \left[1 + \frac{b\sqrt{\pi}}{2\sqrt{c}} e^{b^2/4c} \left\{ \mathrm{erf}(\alpha\sqrt{c}) - 1 \right\} \right]. \quad (16.13)$$

Exercise: Verify equations (16.12) and (16.13).

Some closing comments are in order. In the final model above, the parameter c was taken to be constant. Note that it plays a similar role in equation (16.4) as does b in equation (16.1), notwithstanding the quadratic term: it serves as a measure of the decentralization of the city. As we noted from the tabulated data in Tables 16.1–16.4), b and similarly c in fact tend to decrease over time, corresponding to a "flattening" of the density profile as the city expands radially. This phenomenon may well be a result of improvements in the urban transportation systems (see Chapter 7). Also, Newling's four stages of city development are based on a progression of the parameter β in time from negative to positive values; this may not occur in general and so raises a question as to the efficacy (at least quantitatively) of this sequence of stages. However, the tabulated data indicate that the maximum densities were indeed higher in the nineteenth than the twentieth century, so in that sense, the four stages may well be qualitatively correct.

More realistic and advanced models should (and do) include several other factors that generalize those implicit in the above formulations. One such a factor might be temporal variations in population density, allowing for the possibility of saturation; if so, a logistic model might suffice here. Inclusion of diffusion (spread into neighboring regions) would then have the density satisfying a partial differential equation of the "reaction-diffusion" type, at least in a continuum model (see Adam 2006, chap. 14). A highly populated inner core will give rise to congestion, and this will be proportional to the local density within a given region. Another factor, also neglected here, is the effect of (local) immigration and emigration. One might expect a considerable influx of immigrants in a central district undergoing rapid growth, and the reverse for inner city districts that have become "blighted."

It is also the case that single-center cities are becoming a thing of the past; they are becoming more and more "multi-center" in character. Industrial estates, medical facilities, shopping malls, and business parks are being built (or are relocating) to regions away from the city. This of course creates demand for decentralized travel networks, as has been discussed in Chapter 7. Thus the next step in modeling the growth of cities will be to develop approaches that incorporate the multi-center structure in a way that invites more mathematical analysis.

Chapter 17

THE AXIOMATIC CITY

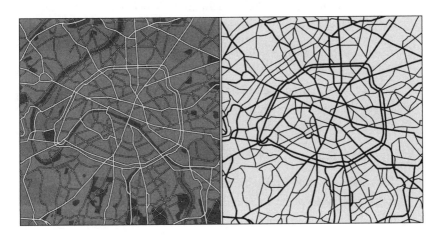

In this chapter we try to be a little more formal by defining axioms for an equilibrium model of a *circular* city. By using the word "equilibrium" here, we mean that the "forces" attracting people to live in the city are balanced at every point by the "forces" that "repel" them. Of course, there are no forces in the physical sense of that word, but by analogy with the balance between gravity and pressure gradients in stars it is possible to suggest certain forms of "coercion" that persuade individuals to live exactly where they do. Let's get started.

$X = \rho(r)$: CIRCLES IN THE CITY

We make the following assumptions (or equivalently, define the following axioms):

I. The population is distributed with circular symmetry in the plane, with radial "civic" population density $\rho(r)$.

II. There is an inward *cohesive force* inducing citizens to relocate nearer the city center; this is because of the desirability of being near work and shopping locations, with lower transportation costs, etc.

III. There is an outward *dispersive force*, specifically a *housing pressure gradient*, inducing citizens to relocate farther from the city center; this is because of higher rental and housing costs in the central regions, etc. This assumption may not be entirely realistic for many modern cities (but we proceed with it nonetheless).

IV. The *civic mass* $M(r)$ is the total amount of "civic matter" interior to r.

V. The *cohesive pressure* $C(r)$ is similar to a two-dimensional "gravitational attraction" in the plane. In a similar manner, it can be written in terms of the civic mass and distance from the city center.

VI. The *housing pressure* $P(r)$ has a power-law dependence on the population density, namely $P(r) = A(\rho(r))^\gamma$ where A and γ are positive constants. (Recall that this power law dependence was mentioned in Chapter 3.)

VII. The city is said to be in equilibrium if at each point within it the two opposing pressure gradients balance one another. This means that there is no net inducement for the populace to move elsewhere (clearly unusual in practice!).

The model we shall develop is essentially a two-dimensional version of the pressure balance equation for stellar equilibrium, as suggested above. Who would have thought that two such different topics could be so intimately connected? Suppressing our astonishment, let's combine these assumptions in mathematical form by referring to Figure 17.1. The civic mass is readily seen to be

$$M(r) = \int_0^r 2\pi\xi\rho(\xi)d\xi, \qquad (17.1)$$

where ξ is a dummy radial variable. Consider a small change δC in C corresponding to a small increment δM in a radial direction; from Figure 17.1 we have that $\delta M \approx \rho\delta r$. Since we are assuming a gravity-type attraction, we posit that

$$\delta C = \frac{kM(r)\delta M}{r^\alpha} \approx \frac{kM(r)\rho(r)\delta r}{r^\alpha}.$$

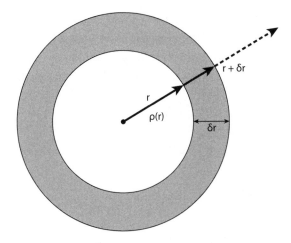

Figure 17.1. Schematic geometry for the civic mass integral in equation (17.1).

In this equation, the "civic coercion" constant k and α are also positive constants; for strictly two-dimensional "gravitational attraction" $\alpha = 1$, but we shall keep it arbitrary for most of this chapter. Assuming further that $C(r)$ is differentiable, we find that, using the standard limiting procedure, that

$$\frac{dC}{dr} = \frac{kM(r)\rho(r)}{r^{\alpha}}. \tag{17.2}$$

The civic pressure gradient is from the higher housing-cost pressures in the central city regions to the lower pressure regions farther out. The cohesive pressure is higher in these external regions, and so the corresponding pressure gradient is in the opposite direction. For a city in equilibrium, therefore, the following equation must be satisfied:

$$\frac{dC}{dr} = -\frac{dP}{dr}, \text{ or } \frac{dC}{dr} + \frac{dP}{dr} = 0. \tag{17.3}$$

From equations (17.2) and (17.3) we find that

$$\frac{kM(r)\rho(r)}{r^{\alpha}} + P'(r) = 0. \tag{17.4}$$

There are several general properties of our "city" that can be established from this equilibrium equation. For example, it follows trivially that the rental gradient $P'(r)$ vanishes at any noncentral location where the population density $\rho(r) = 0$. In particular, the following exercise is left to the reader:

Exercise: Prove the following results:

(a) The central rental gradient $P'(0)$ is always zero at an equilibrium provided $\alpha < 2$. (Hint: use L'Hôpital's rule.)

(b) The quantity $P^+(r) = P(r) + k\frac{M^2(r)}{4\pi r^{\alpha+1}}$ decreases monotonically outward for $r > 0$.

(c) For all r, and $\alpha < 4$, $P(0) - P(r) \geq \frac{kM^2(r)}{4\pi r^{\alpha+1}} \geq 0$.

(d) If the rental $P(r) = 0$ for some radius $r = R$, then the central rental $P(0)$ satisfies $P(0) = \frac{kM^2(R)}{4\pi R^{\alpha+1}} \geq 0$.

Further results can be found by using the following definition of a mean rental $\bar{P}(r)$:

$$\bar{P}(r)M(r) = \int_0^r P(\xi)dM(\xi). \tag{17.5}$$

Exercise: Prove that if the central rental $P(0) \geq 0$, then $\bar{P}(r) - P(r) > \frac{kM^2(r)}{6\pi r^{\alpha+1}} > 0$ and hence that $\bar{P}(r) > \frac{kM^2(R)}{6\pi R^{\alpha+1}}$.

(Hint: integrate by parts twice.)

Now we proceed to look at some more specific city models. We set $\gamma = 2$ in "Axiom" VI and write equation (17.4) in terms of the population density $\rho(r)$ to obtain

$$\pi k \int_0^r \xi\rho(\xi)d\xi + Ar^\alpha \frac{d\rho}{dr} = 0.$$

If we perform a differentiation and define $\beta^2 = A/\pi k$, then we have

$$\rho + \frac{\beta^2}{r}\frac{d}{dr}\left(r^\alpha \frac{d\rho}{dr}\right) = 0.$$

This expression can be rearranged in the form

$$r^2\frac{d^2\rho}{dr^2} + ar\frac{d\rho}{dr} + \frac{r^{3-a}}{\beta^2}\rho = 0. \tag{17.6}$$

This equation is related to the well-known *Bessel equation* of order ν, namely

$$x^2\frac{d^2y}{dx^2} + x\frac{dy}{dx} + (x^2 - \nu^2)y = 0. \tag{17.7}$$

The constant ν may take on any value, but is often an integer. A bounded solution to this equation satisfying the condition $y'(0) = 1$ is $y = J_\nu(x)$, where $J_\nu(x)$ is a Bessel function of the first kind of order ν. Consequently, the corresponding solution of equation (17.6) can be shown to be

$$\rho(r) = \rho(0)r^{(1-a)/2}J_\nu\left(\frac{2}{|3-a|\beta}r^{(3-a)/2}\right), \quad r \geq 0. \tag{17.8}$$

In this expression $\rho(0)$ is the central population density and $\nu = |(1-a)/(3-a)|$. It will be left as an exercise for the reader to verify this result. Note that if $a = 3$, the original differential equation (17.6) reduces to Euler's equation, solutions for which may be found by seeking them in the form $\rho(r) \propto r^m$ and solving the resulting quadratic equation for m. Only solutions for which the real part of m is positive will be appropriate for a city containing the origin, but for an annular city, all solutions are in principle permitted.

Exercise: Verify by direct construction that (17.8) is a solution to (17.6).

(Hint: let $\rho(r) = r^{(1-a)/2}z(r)$ and then make a change of independent variable, $t = r^{(3-a)/2}$.)

Such idealized "cities" can be referred to as "Bessel cities" for obvious reasons. When $a = 1$, the solution (17.8) reduces to the simple form $\rho(r) = \rho(0)J_0(r/\beta)$, and when $a = 2$ the solution is $\rho(r) = \rho(0)r^{-1/2}J_1(2r^{1/2}/\beta)$. From the definition of housing rental the corresponding expressions for $P(r)$ are proportional to the squares of these solutions; they are illustrated in Figure 17.2. The model for $a = 2$ has the disadvantage that, unlike the case for $a = 1$, the gradients $\rho'(0)$ and $P'(0)$ at the center are not zero. While this is not a

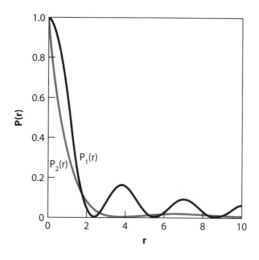

Figure 17.2. $P_1(r)$ (solid line) is the square of the solution (17.8) for $\alpha = 1, \rho_0 = 1$, and $\beta = 1$ (for simplicity). $P_2(r)$ (gray line) is the square of the solution (17.8) for $\alpha = 2, \rho_0 = 1$, and $\beta = 1$.

major problem, this model does not have these "nice" properties enjoyed by the other one.

For Bessel cities, both the density and rental are oscillatory and vanish infinitely often as the distance from the center increases indefinitely. Furthermore, the model predicts that the population density can become negative. Obviously this cannot be the case in reality, so it makes sense to define a *finite Bessel city* by truncating the model at the first zero of the corresponding Bessel function. It is interesting to note that in his $\alpha = 1$ models, Amson (1972, 1973) identifies the positive but ever-decreasing peaks in the Bessel function as "satellite town belts" surrounding a central city, with the regions of negative density identified as greenbelt regions. If the first zero of $J_0(r/\beta)$ occurs at $r = R_1$, the area of this can be called the central area, with central population $M(R_1)$. The zeros of Bessel functions are tabulated and available online. Since the first zero of $J_0(x)$ occurs when $x \approx 2.405$, it follows that $R_1 \approx 2.405\beta = 2.405\,(A/\pi k)^{1/2}$. Hence the central area is

$$\pi R_1^2 \approx 5.783 \frac{A}{\pi k} \propto \frac{A}{k}.$$

From this simple result we see that it depends directly on the rental coefficient A and inversely on the coercion coefficient k, but is independent of the central density $\rho(0)$. It can also be shown that for given values of A and k the central rental $P(0)$ varies as the square of the central population.

Chapter 18

SCALING IN THE CITY

$X = r(n)$: SIZE OF THE CITY

No, by "scaling in the city" we don't mean what Spiderman does in his various movie adventures . . . There is, according to Brakman et al. (2009, available in Oxford's online resource center),

> a remarkable regularity in the distribution of city sizes all over the world, also known as the "Rank-Size Distribution." Take, for example, Amsterdam, the largest city in the Netherlands and give it rank number 1. Then take the second largest city, Rotterdam, and give it rank number 2. Keep on doing this for those cities for which you have data available, possibly selecting only cities exceeding a certain minimum

size. If you calculate the natural logarithm of the rank and of the city size (measured in terms of the number of people) and plot the resulting data in a diagram you will get a remarkable log-linear pattern, this is the Rank-Size Distribution. If the slope of the line equals minus 1, as is for example approximately the case for the USA, India, and France, the relationship is known as Zipf's Law.

This means that the largest city is always about twice as big as the second largest, three times as big as the third largest, and so on, in approximate inverse proportion to its rank. Mathematically, if we rank cities from largest (rank 1) to smallest (rank N) to get the rank $r(n)$ for a city of size n, then $\log r(n) = \log A + a \log n$, or equivalently $r(n) = An^a$, where the parameters A and a are chosen to fit the data. This power law is called an *allometric* relationship (allometry is the study of the change in proportion of parts of an organism as a consequence of growth). Let's elaborate a little on this idea of allometry. If two quantities are related by a power law, $y \propto x^a$ say, then a is called the *scaling exponent*. If $a = 1$ the quantities exhibit *isometric* scaling, for example, change of size without change of proportion (this is also referred to as *geometric similarity*). Allometry is often expressed in terms of a scaling exponent based on the mass M of the object of interest. Thus an isometrically scaling object (such as a cube of side L) would have all volume-based properties change in proportion to L^3, or equivalently as mass to the first power, that is, $V \propto M$; all surface area-based properties change in proportion to L^2, or $M^{2/3}$, and all length-based properties change in proportion to L, or $M^{1/3}$. Another example worth mentioning is *Kleiber's law*: an organism's metabolic rate is proportional to $M^{3/4}$; whereas breathing and heart rates are both proportional to $M^{1/4}$ (see West 1999 for more details).

In 1949, Harvard linguist George Zipf proposed [31] that city sizes (along with many other things) follow a special form of the distribution where $a \approx -1$. This has become known as *Zipf's law*: the frequency of cities within a given size is inversely proportional to their rank. However, as pointed out by Batty *and Longley* [1], most of the work on city-size distributions neglects any spatial structure that exists within cities. By this they mean that cities as measured by populations or incomes, among other measures, are considered as points with their sizes reflecting the competition *between* cities as opposed to competition *within* the city. In other words (and as illustrated by Zipf's law) there are a small number of large cities and a large number of small ones because there are not the resources and demand to sustain many large cities. Zipf's law is empirical in that it has been observed in data taken from a wide range of cities in space and time, but so far no entirely satisfactory theoretical explanation has been found for it. Such an explanation must surely establish the basis for the observed organizational principles that appear to be replicated across such wide spatial and temporal scales.

But it does seem that similar principles must be at work *within* cities, based on microeconomic factors such as population density, rent, employment, transportation costs, and so on. Of interest then, in this context are the constraints that geometry imposes on "density" and "nearness" in a city; a plausible approach by virtue of the fact that sizes of structures indirectly reflect population and employment "volumes." In particular, as buildings grow in size, their shape must change to enable them to function efficiently, and the scaling exponent a is related to the governing allometric "law." Examples include the cost of heating (or cooling) a building; crudely speaking, heat loss will be proportional to the surface area of the building, but the amount of heat required to maintain equable temperatures will depend essentially on the volume of the building. Natural lighting provides another illustration; this will depend on the surface area of the building, but since the area changes with size more slowly than does the volume, the shape of the building must change as it grows to accommodate the requisite increase of natural illumination. In fact, it has been argued that cities yield some the best examples of *fractals* (see below). It is possible to fit power laws and allometric scaling relations to several geometrical properties of buildings—perimeter, area, height, and volume—using a large database of buildings in Greater London, which contains some 3.6 million "building blocks"!

So what might be these basic scaling laws for cities? We mention perhaps just enough to whet the reader's appetite for more advanced discussions of this topic. (The reader may wish to consult Appendix 9 for a short introduction to fractals.) Modifying somewhat the notation in that Appendix to fit the present context, we identify the number of parts composing an object, their total length and area at a given scale a as N, L, and A, respectively. Specifically, they are defined by the relations

$$N(a) = a^{-D}; \; L(a) = N(a)a = a^{1-D}, \text{ and } A(a) = N(a)a^2 = a^{2-D}.$$

TABLE 18.1.
Estimated fractal dimension D for several cities

City	D
Beijing (1981)	1.93
Berlin (1945)	1.69
Boston (1981)	1.69
London (1981)	1.72
Los Angeles (1981)	1.93
Melbourne (1981)	1.85
Mexico City (1981)	1.76
Paris (1981)	1.66
Rome (1981)	1.69
Tokyo (1960)	1.31

If $A(R)$ is the area of the object, considered to be constant regardless of the scale of resolution, then it is reasonable to define the density as

$$\rho(a) = \frac{A(a)}{A(R)} \propto a^{2-D}.$$

If we allow the scale to become finer and finer, so that as $a \to 0$ (assuming $1 < D < 2$), then $N(a) \to \infty$, $L(a) \to \infty$, $A(a) \to 0$, and $\rho(a) \to 0$. In particular, the perimeter becomes infinite as the scale becomes finer. This is an important characteristic (among others) associated with fractal behavior.

In their book *Fractal Cities* [1], Michael Batty and Paul Longley tabulate the estimated fractal dimension D for 28 cities. Table 18.1 contains some of these estimates (to three significant figures). There is also some evidence to suggest that fractal dimension has a tendency to increase as a city grows; Batty and Longley note that from 1820 to 1939 the estimates of D increase from 1.32 to 1.79, though it did drop slightly to 1.77 in 1962 and farther still to 1.72 in 1981.

It will be interesting to see how this relatively new subject of fractal cities itself "evolves" over time. Some indications are already at hand, as discussed in the section below.

$X = N(t)$: LARGER CITIES ARE "FASTER"

In a fascinating paper published in 2007 [32], the first sentence reads "Humanity has just crossed a major landmark in its history with the majority of people now living in cities." The title of this paper, by Bettencourt et al., was "Growth, Innovation, Scaling, and the Pace of Life in Cities." One major finding was that many diverse properties of city life, designated here by $y(t)$, are simple power law functions of population size $N(t)$, that is,

$$y(t) = y_0[N(t)]^\beta. \qquad (18.1)$$

In other words, they "scale" as power laws. Examples of $y(t)$ are total electrical consumption, gross domestic product, and number of gasoline stations. In one sense, this is hardly surprising; these and many other properties will in general increase with population size. Nevertheless, the remarkable fact appears to be that for quantities representing wealth creation and innovation, the power law exponent $\beta \approx 1.2$. $\beta > 1$ implies increasing returns; $\beta < 1$ implies economies of scale. This latter situation occurs for city infrastructure, where $\beta \approx 0.8$. Furthermore, $\beta \approx 1$ is associated with individual human needs (jobs, housing, water consumption), so that these quantities tend to be directly proportional to population size.

Understanding the implications of these power laws is crucial because, as the authors point out, "*a major challenge worldwide is to understand and predict how changes in social organization and dynamics resulting from urbanization will impact the interactions between nature and society.*" There is a balance between the innovative and destructive aspects of city living, however. In addition to providing large-scale social services, education, and health care, for example, cities are the main source of crime, pollution, and disease in society.

In some ways cities can be thought of as living organisms because they consume energy and produce artifacts and waste. It is therefore not surprising that there are biological metaphors associated with the "scaling" phenomenon illustrated above (as already noted in Chapter 1). For example, as indicated above, the surface area of an object is proportional to (i.e., scales as) the square of its size (linear dimension), and its volume scales as the cube of its size. More specifically, many if not most physiological properties of biological organisms scale with body mass M with an exponent β that is a multiple of ¼ (this is a generalization of Kleiber's law, mentioned above). Thus the metabolic rate per unit mass $R \propto M^{-1/4}$ decreases with body size. This means that larger organisms consume less energy per unit time and per unit mass. Inversely, life spans and maturation times scale as $M^{1/4}$. Nevertheless, there are differences: larger organisms live by slower biological clocks, by virtue of the inverse quarter power law referred to above, whereas the pace of city life *increases* with size, as we shall see below. As the authors of the article point out, scaling has proved to be a valuable tool for revealing underlying dynamics and structure for many problems encountered in science and technology. In what follows, we discuss some implications of the "urban growth equation" derived in that paper.

We are interested in deriving and subsequently solving an equation for the population growth rate, dN/dt, in terms of resources for maintenance and growth per unit time mentioned above (y) and two other parameters, R and E. R is defined as the amount of resource(s) per unit time required to maintain an individual; E is the amount required to add another individual to the population. Then for ΔN individuals added in time Δt the resource allocation rate equation may be written, in the appropriate limit as

$$y = RN + E\frac{dN}{dt}. \tag{18.2}$$

The left-hand side of this equation is a measure of the available resources, balanced on the right by consumption terms due to maintenance and population growth, respectively. Substituting for y from equation (18.1) and rearranging we find that

$$\frac{dN}{dt} + \left(\frac{R}{E}\right)N(t) = \left(\frac{y_0}{E}\right)[N(t)]^{\beta}. \tag{18.3}$$

This is a *Bernoulli equation*; the substitution $v = N^{1-\beta}$ reduces this to a linear equation with integrating factor

$$\exp\left(\frac{R(1-\beta)t}{E}\right).$$

The resulting solution of equation (18.3) for $\beta \neq 1$ is

$$N(t) = \left[\frac{y_0}{R} + \left(N_0^{1-\beta} - \frac{y_0}{R}\right)\exp\left(-\frac{R}{E}(1-\beta)t\right)\right]^{\frac{1}{1-\beta}}, \tag{18.4}$$

where $N_0 = N(0)$.

Exercise: Derive equation (18.4).

There are three basic forms of solution depending on whether $\beta > 1$, $\beta = 1$ or $\beta < 1$. The simplest case, $\beta = 1$, has the solution (using equations 18.1 and 18.2)

$$N(t) = N_0 e^{(y_0 - R)t/E}. \tag{18.5}$$

This represents, of course, simple exponential growth or decay depending on the sign of the exponent. For $\beta < 1$ the growth curve is sigmoidal (sometimes called Gompertz-like); as can be seen from equation (18.4), there is a horizontal asymptote—the carrying capacity (see Chapter 15)—found by taking the limit as $t \to \infty$. Its value is $N_\infty = (y_0/R)^{1/(1-\beta)}$. This is quite a significant result, for it means that there is an upper limit to the size of the population: it will eventually stop growing (in practical terms, after a very long but finite time). As we saw in Chapter 15, this is very similar to biological systems in which the competition between finite resources and population growth eventually results in saturation of the population (even in the simplest models). The urban equivalent of "finite resources" in biology or ecology is referred to in the article as "economies of scale." A typical graph of $N(t)$ is shown in Figure 18.1 for a simple choice of parameters in equation (18.4), expressing in units of, say, 100,000. The initial value $N_0 = 1$, with $\beta = 0.5$, and the carrying capacity $N_\infty = 25$ are chosen here for illustrative purposes.

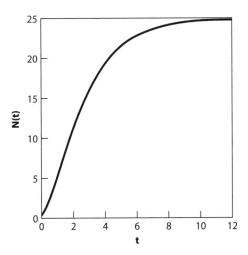

Figure 18.1. Typical solution $N(t)$ based on equation (18.4).

The case of $\beta > 1$ is even more interesting. The solution can be rearranged in the form

$$N(t) = \left\{ \left[\frac{y_0}{R} + \left(\frac{1}{N_0^{\beta-1}} - \frac{y_0}{R} \right) \exp\left(\frac{R}{E}(\beta-1)t \right) \right]^{\frac{1}{\beta-1}} \right\}^{-1}. \quad (18.6)$$

If $N_0 > (R/y_0)^{1/(\beta-1)}$, then the term inside the curly brackets will become zero for some time t_c; it corresponds to unbounded (or explosive) growth at finite time (and therefore the growth is faster than exponential). In fact, this condition on N_0 is readily obtained from equation (18.3). We have previously encountered this kind of growth, albeit in a less sophisticated form, in Chapter 15.

Since equation (18.6) is of the form $N(t) = [A - Be^{Ct}]^{-D}$ for positive constants A, B, C, and D, it follows that

$$t_c = \frac{1}{C}\ln\left(\frac{A}{B}\right) = -\frac{E}{R(\beta-1)}\ln\left(1 - \frac{RN_0^{1-\beta}}{y_0}\right). \quad (18.7)$$

Noting that $\ln(1-x) \approx -|x|$ for $|x| \ll 1$, we see from the above equation that, provided the quantity $RN_0^{1-\beta}/y_0$ is small "enough" compared with 1, then

$$t_c \approx \frac{E}{y_0(\beta-1)N_0^{\beta-1}}. \quad (18.8)$$

A graph of $N(t)$ in this case showing the (theoretically) unbounded growth is shown in Figure 18.2. Of course, for a finite city size (or population) and finite resources, such a singularity cannot occur. However, the authors point out that left unchecked, this lack of sustainability triggers a "phase transition" to the point where $N(t)$ collapses; according to equation (18.3) this will occur when N exceeds the critical value $N_c(t) = (R/y_0)^{1/(\beta-1)}$ (this was the value of N_∞ for $\beta < 1$). At this point $N'(t) < 0$ and an inexorable population decline follows (see Figure 18.3) unless, as the authors point out, major qualitative changes occur which effectively "reset" the initial conditions and parameters of the governing solution defined by equation (18.4). In order to maintain growth, then, a new cycle must be initiated wherein $\beta > 1$ as before, and at the "new" time $t = 0$ the parameters N_0, R, and y_0 are such that, as before, $N_0 > (R/y_0)^{1/(\beta-1)}$. While in principle this process can be repeated continually, leading to multiple cycles (and pushing potential collapse into the future), there is an inherent problem. Note from equation (18.8) that the time between cycles t_c is inversely proportional to $N_0^{\beta-1}$. Clearly this decreases with each new cycle, since N_0 increases

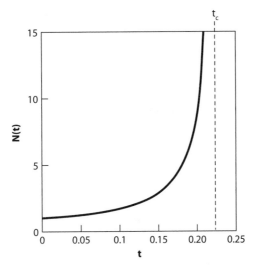

Figure 18.2. Typical super-exponential growth with finite-time singularity t_c (vertical line).

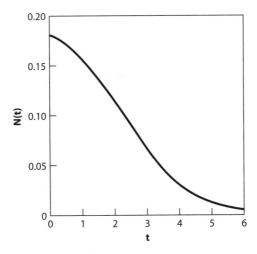

Figure 18.3. Population collapse when growth cannot be maintained.

accordingly. Therefore major innovations and adaptations must occur at an accelerated rate; in other words, the pace of life increases with city size.

At this point the reader may wonder yet again whether this scenario has any basis in reality. Citing several reputable sources, the authors note that "these predicted accelerating cycles . . . are consistent with observations for the population of cities . . . , waves of technological change, and the world population." And to some degree these ideas can be quantified: the famous Scottish scientist Lord Kelvin is reputed to have said that "I have no satisfaction in formulas unless I feel their arithmetical magnitude," so let's try to do just that. In equation (18.8) the ratio E/y_0 appears. This may be interpreted as the time required for an average individual to reach "productive maturity," or more simply, the time needed to "create" a new individual. Bettencourt et al. [32] express this as $E/y_0 \sim (20 \text{ yr})T$, T being a number of order unity. With $\beta = 1.1$ and the initial population measured in millions ($N_0 = 10^6 n$), equation (18.8) reduces to the simple form

$$t_c \approx 50\left(\frac{T}{n^{0.1}}\right) \text{ yrs.} \tag{18.9}$$

For a large city t_c will typically be a few decades, but clearly this timescale decreases with increasing city size. This transition to successive cycles of super-exponential growth accompanied by a reduction of t_c has, it seems, been a common pattern in urban development [29] as well as for the world population [33]. As an example of this urban pattern, Batty and Longley [1] consider the population growth of the New York metropolitan area from 1790 to the present day. It can be decomposed into successive periods of super-exponential growth, and the period of faster growth in the 1960s was followed by the decline of the 1970s as individuals left the city under "the perception of spiraling increases in costs, crime and congestion."

But super-exponential growth, you ask? Are we back at the Doomsday equation once more? We can do no better than to quote again from Johansen and Sornette [29]:

> It is estimated that 2000 years ago the population of the world was approximately 300 million and for a long time the world population did not grow significantly, since periods of growth were followed by periods of decline. It took more than 1600 years for the world population to double to 600 million and since then the growth has accelerated. It reached 1 billion in 1804 (204 years later), 2 billion in 1927 (123 years later), 3 billion in 1960 (33 years later), 4 billion in 1974 (14 years later), 5 billion in 1987 (13 years later) and 6 billion in 1999 (12 years later). This rapidly accelerating growth has raised sincere

worries about its sustainability as well as concerns that we humans as a result might cause severe and irreversible damage to eco-systems, global weather systems etc.

The interested reader is recommended to consult this rather technical paper for further details of the suggested "finite-time singularity" and its possible manifestations and consequences for humanity. Related, but much earlier papers (including von Foerster et al. [28]) are also well worth reading; the reader should also consult the additional references for details.

Chapter 19

AIR POLLUTION IN THE CITY

I spent several years as a student living in London, but fortunately I never had to experience something that plagued the city in the first half of the twentieth century: smog (= smoke + fog). The last major occurrence of London smog was in 1952, and while estimates vary, it is thought that as many as 12,000 people died in the weeks and months following the outbreak. Basically, smog is caused by the chemical reaction of sunlight with chemicals in the atmosphere.

But more generally, what is air pollution? Essentially, it is the presence of substances in the atmosphere that can adversely affect the quality of human, animal, and plant life, and the environment. Of course, this is rather vague, and the definition itself is rather "fluid," changing somewhat as more is known about pollutants. Furthermore, it is usually the addition of such substances resulting directly or indirectly from human activity that is of most concern. The biggest such contributions are from the burning of fossil fuels, including coal, oil, and gas in cars, trucks, factories, and homes. However, natural sources such as forest fires and volcanic eruptions can cause local and even global havoc, as was seen in April 2010 when the Icelandic volcano *Eyjafjallajökull* erupted.

Air pollution is designated primary or secondary. The former results from pollution that is introduced directly into the air, such as smoke and car exhausts. Secondary pollution forms in the air as a result of sunlight-induced chemical reactions changing the nature of the primary pollutants.

$X = V_t$: PARTICLES IN THE CITY

Let's examine the behavior of small particles like aerosols or tiny cloud droplets as they fall slowly through air (or indeed, sediment as it settles down to the bottom of a lake). Both are well described by *Stokes' law*, stated below, provided their speed of descent V is small enough that no turbulence is generated in their wake. Not surprisingly, the upper limit for aerosol sizes is determined by sedimentation—unless the particles can stay aloft for reasonable periods of time (days or longer), they will contribute little to the lack of long-term visibility.

Consider for simplicity a spherical particle of radius R and density ρ falling through the air. The downward force acting on the particle is its weight,

$$W = \frac{4\pi}{3} R^3 \rho g, \qquad (19.1)$$

g being the gravitational acceleration. (Although forces are vector quantities, all the ones acting on this particle are directed upward or downward, so the vector notation will be dispensed with here.) There is an upward buoyancy force, but since the density of the air is so small compared with that of the particle, it will also be neglected here. The other force acting to resist the downward fall of the particle is the drag force F; as the particle initially accelerates downward because of gravity, this resistive force will also increase until the two are in balance, assuming the time of fall is long enough to permit this, as it will usually be in this context. Then there is no net force acting on the particle, and it proceeds to fall with a constant speed, the *terminal speed*. (The correct term is terminal *velocity*, but again, we are not concerned with vector notation here.)

Naturally, the magnitude of the terminal speed will determine how long the particle how long the particle remains in the air. Very small particles ($< 0.1\mu$) are continually buffeted by a molecular process known as *Brownian motion*, and remain suspended indefinitely.

A simple argument can be used to determine the dependence of the drag force on the particle size and speed V. It is reasonable to assume that $F \propto \eta$, η being the (dynamic) viscosity coefficient of the air, that is, $F = K\eta$, where K is a (dimensional) constant. If we equate the dimensions of both sides of this equation using $[M], [L]$, and $[T]$ for mass, length, and time, respectively, then

$$[M][L][T]^{-2} = [\text{dimensions of } K] \times [M][L]^{-1}[T]^{-1}.$$

Therefore the dimensions of K are $[L]^2[T]^{-1}$ and this is accomplished, in particular, by the combination RV. The magnitude of K cannot be determined by dimensional arguments, but in 1851 George Gabriel Stokes carried out more detailed calculations and found that for small Reynolds numbers R (where $R = VR\rho_a/\eta < 1$, ρ_a being the density of the air),

$$F = 6\pi\eta RV. \tag{19.2}$$

(The reader may recall that earlier in this book (Chapter 3) a more generic form of the Reynolds number was used, namely $R = ul/\nu$, where u, l, and ν represented a typical speed, length scale, and (kinematic) viscosity, respectively. To avoid confusion here with the radius R, we use the alternative notation Re for the Reynolds number. Also, since this chapter is a little more technical and mathematically precise, the latter form for the Reynolds number is more appropriate here (note also that $\nu = \eta/\rho_a$).)

To determine the dependence of the terminal speed V_t on the particle size, we just equate the two opposing forces and solve equation (19.2) for this speed to find that

$$V_t = \frac{2\rho g}{9\eta}R^2. \tag{19.3}$$

Clearly, the terminal speed increases quite rapidly with particle size. Let us apply this result to calculate the terminal speed of a particle of radius 10 microns (10^{-5} m) in still air at 5°C at an altitude of 1000 m; $g \approx 9.8$ m/s², $\rho \approx 2 \times 10^3$ kg/m³ and $\eta \approx 1.8 \times 10^5$ Ns/m². Therefore

$$V_t = \frac{2 \times 2 \times 10^3 \times 9.8 \times (10^{-5})^2}{9 \times 1.8 \times 10^{-5}} \approx 2.4 \times 10^{-2} \text{ m/s},$$

or about 2 cm/s. The Reynolds number for this particle in air is, from the definition above with $\rho_a \approx 1.1 \text{ kg/m}^3$,

$$\text{Re} \approx \frac{2.4 \times 10^{-2} \times 10^{-5}}{1.8 \times 10^{-5}} \approx 1.4 \times 10^{-2} \ll 1,$$

so the use of Stokes' law is readily validated. Note that a particle ten times smaller $(R = 1\mu)$ falls one hundred times more slowly, at about 2×10^{-4} m/s. In Chapter 20 these ideas of sedimentation rates will be discussed again, but in the context of reduced visibility caused by particulate scattering of light.

$X = C(x,t)$: POLLUTION IN THE CITY

How do the suspended particles spread in the air? In the absence of wind and other air currents the simplest approach to this question is to consider the one-dimensional diffusion equation derived in Appendix 10:

$$\frac{\partial C}{\partial t} = D \frac{\partial^2 C}{\partial x^2}. \tag{19.4}$$

For simplicity we have assumed that the diffusion coefficient D is constant (although there are certainly situations where this is not the case). This type of argument can be readily generalized to the case of two or three dimensions and geometries other than Cartesian, and we will just introduce them as needed from this point on. This equation describes a trend to a uniform distribution of the pollutant concentration $C(x,t)$ over time. To see this, suppose that for a particular interval of time, $C(x,t)$ has a local minimum; then the right-hand side of equation (19.4) is positive, and C will increase in time. Correspondingly, if C has a local maximum, C will decrease in time.

There is another mechanism that must be included in any realistic discussion of pollution: wind. We shall consider the effects of a wind with constant speed U in the x-direction only (even when a higher-dimensional diffusion equation is used). Again, as shown in Appendix 10, equation (19.4) can be generalized to become

$$\frac{\partial C}{\partial t} = D\frac{\partial^2 C}{\partial x^2} - U\frac{\partial C}{\partial x}. \qquad (19.5)$$

In most cases of interest, this new wind "advection" term far outweighs the effects of the diffusion term. A more practical variant of the problem for our purposes is represented by the equation

$$\frac{\partial C}{\partial t} = D\left(\frac{\partial^2 C}{\partial y^2} + \frac{\partial^2 C}{\partial z^2}\right) - U\frac{\partial C}{\partial x}. \qquad (19.6)$$

What, then, does this equation signify? In this context, it describes the temporal and spatial behavior of the concentration of pollutant particles as they diffuse in the y-z plane perpendicular to the wind direction (x) wafting them downstream. A further simplification is often justified, namely, to consider a steady-state situation. Frequently the source of pollutants emits them at a constant rate, and has a plume whose average shape doesn't change much in time, unless the wind direction changes or some new weather pattern otherwise modifies it significantly. If these do not occur, then we may set the left-hand side of equation (19.6) to zero, resulting in an impressive-sounding (but less impressive-looking) time-independent advection-diffusion equation! Here it is:

$$U\frac{\partial C}{\partial x} = D\left(\frac{\partial^2 C}{\partial y^2} + \frac{\partial^2 C}{\partial z^2}\right). \qquad (19.7)$$

We proceed to justify a solution of this equation in a nonrigorous way as follows.

Exercise: Show by direct substitution that the equation

$$\frac{\partial C}{\partial x} = D\frac{\partial^2 C}{\partial y^2} \qquad (19.8)$$

possesses a solution for $x > 0$

$$C(x,y) = \frac{K_1}{\sqrt{x}}\exp\left(-\frac{y^2}{4Dx}\right). \qquad (19.9)$$

The source is a "point" located at $x = 0$, $y = 0$, but do not be concerned about the apparent (and real) singular behavior of this solution there; textbooks on partial differential equations discuss this type of problem and its resolution (with time t replacing x). K_1 is a constant that depends on the rate of pollutant emitted in the plane $x = 0$, and possibly on D and U (see below).

Because of the symmetry in y and z possessed by equation (19.7), and based on the solution (19.9), we expect the solution of (19.7) to be of the form

$$C(x,y,z) = \frac{K_2}{x} \exp\left(-\frac{(y^2 + z^2)U}{4Dx}\right). \tag{19.10}$$

This is also readily verified by direct substitution (exercise!). Generally, K_2 will also depend on the constants D and U. (We have simplified things greatly here; generally D will be different in the horizontal y-direction, perpendicular to the wind, from that in the vertical z-direction.) Note that, for a given distance x downstream, the maximum concentration is $C(x,0,0) = K_2(D,U)/x$.

Now let's think "laterally" for a moment; the right-hand side of equation (19.9) looks suspiciously like the equation of the *normal distribution*. (I have a now-defunct ten Deutschmark bill with the equation and graph of that distribution on it.) This has the classic "bell-curve" shape. In statistics, the random variable X is said to be normally distributed with mean μ and standard deviation σ if its probability distribution is given by

$$f(X) = \frac{1}{\sigma\sqrt{2\pi}} \exp\left[-\frac{(X-\mu)^2}{2\sigma^2}\right]. \tag{19.11}$$

The total area between the X-axis and the graph of the function $f(X)$ is equal to one (see Figure 19.1). If we compare equation (19.11) for f with that for $C(x,y)$ above in equation (19.9), it is clear that $\mu = 0$, $\sigma = \sqrt{2Dx}$, and $K_1 = 1/\sqrt{4\pi D}$. Carrying this equivalence over to the solution (19.10) for $C(x,y,z)$, we see that the plume concentration is distributed normally in the crosswise direction (y) but also vertically (z), because now $\sigma = \sqrt{2Dx/U}$ and $K_2 = 1/\sqrt{4\pi D/U}$. The standard deviation σ is a measure of how tightly the concentration sits about the mean value μ (zero here); so the plume width in the x- and z-directions is here proportional to $\sqrt{x}$, the square root of the distance downstream. It is also proportional to the square root of the diffusion coefficient and inversely proportional to the square root of the wind speed U. It is known that turbulence can increase D (which in this context is often referred to as the eddy diffusivity).

In light of these comments, let us summarize the predictions of the above model based on the solution (19.10) for $C(x,y,z)$ now dropping the subscript on K_2:

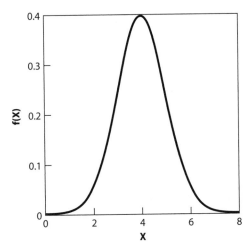

Figure 19.1. Normal distribution $f(X)$ given by equation (19.11) for $\mu = 4$ and $\sigma = 4$.

I. The downwind concentration is directly proportional to $K(D, U)$, the source emission rate.

II. The more turbulent the atmosphere, the wider the lateral spread of the plume after any given time.

III. The maximum concentration at any location is found at ground level on the line $y = 0$, and is inversely proportional to distance x downwind from the source.

A more general analysis of the particle emission problem shows that $K \propto U^{-1}$, so that the maximum concentration will be smaller for higher wind speeds. But even a less sophisticated model such as ours can be adapted to emission from an elevated source. A common example is the plume of smoke from a smokestack. Because the particles have a longer time to diffuse before they reach the ground, we expect that the maximum concentration will depend on the height H of the stack. Let's investigate this. The simplest and most obvious modification to the model is to replace z by $z - H$ in equation (19.10) for $C(x,y,z)$. However; there is now a lower boundary for the particles—the ground—and the associated problem of what boundary condition to impose on the problem. At this point the reader may object: in the previous model the source is on the ground at "height" $z = 0$; so why worry about the effect of ground in this model? In that case the maximum concentration was always at ground level, so the effect was implicit in the model. As we shall see, a simple modification is to assume that particles are partially reflected from the ground when they diffuse and settle downward. Perfect reflection is unlikely, of course; the ground is not a mirror and the particles are not perfectly elastic—indeed, recalling the difference that clay or grass courts can make in tennis, the problem is rather more complex than we can investigate in depth here. Nevertheless, some insight into the modeling process and the conclusions drawn from it can be illustrated by assuming that there is an "image" smokestack emitting pollutants at a rate $\alpha K (0 \leq \alpha \leq 1)$ from $z = -H$. Of course, this is just a mathematical artifact (and a common one in this kind of problem). We are only interested in what happens for $z \geq 0$. To this end, the suggested form for the concentration distribution is

$$
\begin{aligned}
C(x,y,z) = \frac{K}{(1+\alpha)x} &\exp\left(-\frac{y^2 U}{4Dx}\right) \left[\exp\left(-\frac{[(z-H)^2]U}{4Dx}\right) \right. \\
&+ \left. \alpha \exp\left(-\frac{[(z+H)^2]U}{4Dx}\right) \right].
\end{aligned}
\tag{19.12}
$$

The divisor $1 + \alpha$ in this expression is required to ensure that C reduces to the case above when $H = 0$. Again, we are interested in the concentration at ground level, which simplifies to

$$
C(x,y,0) = \frac{K}{x} \exp\left(-\frac{(y^2 + H^2)U}{4Dx}\right).
\tag{19.13}
$$

For any given $x > 0$ this will have its maximum value when $y = 0$, so it will be of interest to determine the location of the maximum of the function

$$
C(x,0,0) = \frac{K}{x} \exp\left(-\frac{H^2 U}{4Dx}\right).
\tag{19.14}
$$

Using the first derivative test it is straightforward to show that the maximum ground-level concentration of

$$
C_{max} = \frac{4DK}{eUH^2}
$$

at $x = x_m = UH^2/4D$. Using this very simple model we have been able to reproduce a result first derived in 1936 [34], namely that the maximum ground-level concentration from a plume released from a height H is inversely proportional to H^2.

Exercise: Verify this result.

Summarizing, therefore, the model predicts that

I. C_{max} is inversely proportional to the square of the plume-release height H.

II. C_{max} is inversely proportional to the wind speed U.

III. x_m is directly proportional to the square of the plume-release height H.

IV. x_m is directly proportional to the wind speed U.

V. C_{max} and x_m are both independent of the ground "reflection coefficient" α.

The model developed here is very simplistic, so it is encouraging to note that the first two predictions are the same as those from more sophisticated models. Physically, they make sense, since the plume is diffusing in both the y- and z-directions as the wind carries it downstream, and the pollutants are spread over a wider area (which has dimensions of $(\text{length})^2$). Furthermore, a stronger wind will stretch out the plume more per unit time, diluting it all the more as it does so.

As for the remaining predictions, III and IV follow naturally for the same reasons as I and II. In reality, the effect of ground reflection must play a role, though perhaps only a minor one compared to that of H and U. One mechanism neglected here is that of *buoyancy*; very often the effluents released (or, indeed, smoke from forest fires) is warmer than the surrounding air, and it continues to rise for a time after it is released. But to keep things relatively simple, that effect has not been included here.

$X = C(x,t)$: A DISTRIBUTED SOURCE

We have regarded the source of effluent to be on the ground or at the top of a smokestack. In each case the source is in effect a *point source*. This is because of the nature of the solution at $x = 0$, referred to above. But consider a long line of slowly moving bumper-to-bumper traffic along a straight stretch of road. This can be considered a distributed source of particulates (current emission regulations notwithstanding)—a *line source*. Of course, the average speed of the traffic, the length of the road, and the wind strength and direction will affect the concentration of particles (such as hydrocarbons from the tailpipes) at any point on or off the road. To build a simple model of pollutant dispersal for a line of traffic of length L, we now use the emission rate per unit length, namely K/L. We will again neglect buoyancy and regard the line source as being placed at ground level along the y-axis from $-L/2$ to $L/2$ (though it is not necessary to specify this in what follows). We shall utilize the earlier models by considering only a cross-wind U in the x-direction as before; thus the pollutant is blown directly from the road into the neighboring land or cityscape. For a long traffic line L (strictly, an infinitely long line) there can be no variation of C in the y-direction because the source is uniform along that line. Therefore we approximate the finite-L case by requiring that the particles diffuse only in the vertical direction (again, the effect of wind dominates any diffusion in the x-direction). The governing equation now simplifies to

$$U\frac{\partial C}{\partial x} = D\frac{\partial^2 C}{\partial z^2}.\qquad(19.15)$$

Again, dropping the subscript, this time on K_1, the solution, based on equation (19.9), is now

$$C(x,z) = \frac{K}{L\sqrt{x}}\exp\left(-\frac{z^2}{4Dx}\right).\qquad(19.16)$$

Note that at ground level ($z = 0$) the concentration varies more slowly downwind, as $x^{-1/2}$, compared with x^{-1} for a point source. Clearly the concentration for any $x > 0$ is maximized at ground level, but this result is a means to an end. In many situations, the source will be better approximated, not by a point or a line, but by an *area* composed of multiple sources in an urban region. These can effectively combine because of wind and diffusion in such a way as to render the individual sources unidentifiable. Since we are interested in the ground-level concentration, we set $z = 0$ in the above equation and imagine for simplicity a *rectangular* source by integrating the result with respect to x. Therefore the accumulated concentration $C_A(x)$ has the following dependence on x:

$$C_A(x) \propto \int_0^x C(\xi, 0)\, d\xi \propto \lim_{b \to 0^+} \int_b^x \xi^{-1/2} d\xi \propto \sqrt{x}. \qquad (19.17)$$

More realistic models indicate a greater dependence on x than this, especially if the atmospheric conditions preclude the particles from unlimited diffusion vertically. There is evidence to suggest that the rate at which pollutants are emitted and the region affected by pollution both increase faster than the population does. Modeling this would be a very substantial exercise for the reader!

Chapter 20

LIGHT IN THE CITY

With such particles suspended in the atmosphere for sometimes days or weeks at a time, smog presents a danger to health, but in London it was also known as a "pea souper" because one could not see one's hand in front of one's face! In fact, as a result of the Great London Smog of 1952 (caused by the smoke from millions of chimneys combined with the mists and fogs of the Thames valley), the Clean Air Act of 1956 was enacted. With this in mind we now turn to the topic of how air pollution may affect *visibility*.

$X = I_s$: VISIBILITY IN THE CITY

We start with an apparent *non sequitur* by asking the following question. Have you ever been in an auditorium of some kind, or a church, in which your view of a speaker is blocked by a pillar, but you can still hear what is being said? I'm pretty sure you must have experienced this. Why can your ears receive auditory signals, but your eyes cannot receive direct visual ones (excluding Superman of course)? The reason for this is related to the wavelengths of the sound and light waves, being ≈ 1m and $\approx 5 \times 10^{-7}$ m, respectively. The latter, in effect, "scatter" more like particles while the former are able to diffract ("bend") around an obstacle comparable in size to their wavelength. By the same token, therefore, we would expect that light waves can diffract around appropriately smaller obstacles, and indeed this is the case, as evidenced by softly colored rings of light around the moon (coronae) as thin cloud scuds past its face. Another diffraction-induced meteorological phenomenon is the green, purple-red, or blue iridescence occasionally visible in clouds. But it is the collective phenomenon known as *scattering* that we wish to discuss in some detail, in order to better appreciate the character of air pollution and its effect on the light that reaches our eyes.

So: when light is deflected in some manner from its direction of travel, it is said to be scattered. There are several mechanisms that contribute to the scattering of light by particles in the atmosphere: reflection, refraction, and diffraction being the most common, though they are not necessarily mutually exclusive effects. The size of the particles determines which mechanism is the predominant one.

Visibility is reduced to some extent by the absorption of light, but scattering by particles and droplets is the primary source of this reduction. We perceive distant objects by contrast with their environment, and this contrast is reduced by the scattering of light from particles and droplets in the line of sight. Thus visibility is reduced. Depending on their size, particles may settle out of the air in due course; this sedimentation process had already been mentioned in the previous chapter.

Although the particles will be irregular in shape, we can define an effective radius as the average of that of (i) the largest sphere that can be inscribed in the particle and (ii) the smallest sphere that contains the particle. If this radius exceeds about 10 microns (10^{-5} m), they settle out in several hours. As we saw, when an object falls in air, it is subject to at least two separate forces; its weight (acting downward) and air resistance or drag (acting upward). A third force is that of buoyancy (also acting upward), but since the air density is negligible compared with that of the particle, this can be neglected. The particle weight is proportional to its mass and hence to the cube of its radius. As we saw in the previous chapter, the air resistance is only proportional to its radius, so the weight dominates the drag by a factor that increases as the square of the radius; hence larger particles fall faster, at least initially. As the speed increases, so does the air resistance. If the particle falls from a sufficient height, these competing forces eventually balance each other, and the net force is zero. At this point the particle falls with a constant speed, the terminal speed. On the other hand, if the particles are smaller than 10 microns in size, they can remain suspended in the air for several days, buffeted by air currents.

This, then, is a qualitative summary of the hydrodynamic aspects of sedimentation discussed earlier. By contrast, the optical aspects are more complicated because of the range of particle sizes compared with the wavelengths of visible light (approximately $0.4–0.7\mu$) A convenient measure of relative size is the radius-to-wavelength ratio R/λ. When this ratio is at least about ten, the particles are considered to be large, and it is convenient to regard light in terms of rays. This is the domain of *geometrical optics*, and as illustrated in Figure 20.1, the three processes mentioned above can occur. Light rays can be partially reflected from the surface of the particles, refracted on passing through the interior, or diffracted ("bent") around the edges. All three mechanisms are exhibited in the phenomenon of the rainbow (see Appendix 11); light is refracted and reflected by raindrops to produce this beautiful colored arc in the sky. Less familiar is the third important mechanism—diffraction—a consequence of the wavelike properties of light. This is responsible for some of the more subtle rainbow features—pale fringes below the top of the bow and, as already noted, iridescence in clouds near the sun.

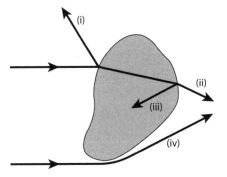

Figure 20.1. Light incident on a large particle may undergo some or all of the indicated processes: (i) reflection at the surface; (ii) refraction into and out of the particle; (iii) internal reflection; and (iv) edge diffraction. Redrawn from Williamson (1973).

For large particles, the amount of light "scattered" by diffraction is as much as that by the other two mechanisms. Some of the refracted light may be absorbed by the particles; if so, this will affect the color of the outgoing radiation. An extreme example of this is black smoke—in this case most of the incident radiation is absorbed. When little or no absorption occurs, large particles scatter light pretty much in the forward direction, so the observer looking toward the light source—the sun, usually—will see a general whitish color. When the particles are not large, the "light ray" approach of geometrical optics is inadequate to describe the scattering processes; the wave nature of light, as mentioned above, must be taken into account. Particles for which $R/\lambda \approx 1$ scatter light in a wider "band" away from the incident direction (resulting in the sky appearing hazy), but smaller particles (for which $R/\lambda \ll 1$) are better able to scatter light multidirectionally. If the aerosols are smaller than about $0.1\,\mu$, the light is scattered much more uniformly in all directions; as much backward as forward and not much less off to the sides. Furthermore, the amount of scattering is very sensitive to the wavelength of the incident light; as we will see below the degree of scattering is $\propto \lambda^{-4}$. This means that the light of shorter wavelength, such as blue or violet, is scattered much more than the longer wavelength red light. Only when we look in the direction of the setting sun, for example, do we see the red light predominating—most of the blue has been scattered out of the line of sight. Think for a moment of cigarette smoke curling upward from an ashtray; typically it is bluish in color—a consequence of the smoke particles being smaller than the wavelengths of light. It is the blue light that is scattered more, and this is what we see.* This is an example of *Rayleigh scattering*, the same phenomenon that makes the sky blue. Rayleigh scattering arises because of wavelength-dependent molecular scattering

To understand the phenomenon of scattering from a more analytic point of view, we need to recall some basic physics. An electromagnetic wave has, not surprisingly, both an electric and a magnetic field that are functions of time and space as it propagates. The direction of propagation and the directions of these fields form a mutually orthogonal triad (Figure 20.2). And when an electromagnetic field encounters an electron bound to a molecule, the electron is

---

* But note that , if the smoke is exhaled, it appears to be whiter because moisture from the air in the lungs has coated the smoke particles. This increases their effective size, and alters the wavelength dependence of the scattering. Light is now scattered more uniformly than before, and hence the smoke appears whiter.

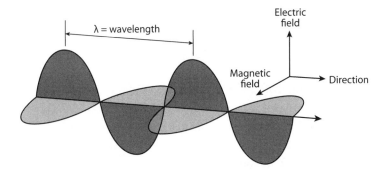

Figure 20.2. Orthogonal triad formed by the direction of the electric and magnetic fields and the direction of propagation for an electromagnetic wave.

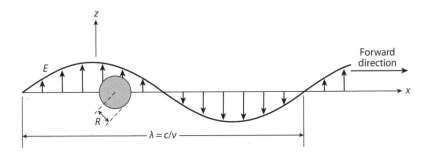

Figure 20.3. A spherical particle of radius $R \ll \lambda$ experiences a nearly constant electric field E. (The wavelength of the incident light is λ.) Redrawn from Williamson (1973).

accelerated by the electric field of the wave. It's a type of "chicken and egg" situation, because an accelerated electron will also radiate electromagnetic energy in the form of waves in all directions (to some extent), and this is the scattered radiation that we have been discussing. Consider Figure 20.3, which illustrates such a situation for a small particle with $R/\lambda \ll 1$ as a snapshot in time.

As shown in the figures, the wave propagates with speed c in the x-direction, with the electric field in the z-direction (it is said to be *polarized* in that direction; this is a qualification we shall address below). This field varies periodically with frequency, $\nu = c/\lambda$ (where λ is the wavelength), and its fluctuations affect the electrons it encounters. To a much lesser extent, the more massive nuclei are also affected, but this will be ignored here. We shall denote the incident electric field by

$$E(x_0, t) = E_0 \sin \omega t, \tag{20.1}$$

where x_0 is the location of this particular electron on the x-axis, E_0 is the amplitude of the wave, and $\omega = 2\pi\nu$ is called the angular frequency of the wave. As the wave passes, electrons will be accelerated back and forth in the z-direction, which in turn will radiate electromagnetic waves—this radiation is the scattered light. An important consequence of our assumption that the particle is small is this: since $R \ll \lambda$ the electric field is almost uniform throughout the particle, so every electron (with charge e) experiences close to the same force (eE) accelerating it, proportional to its displacement s from its former position of equilibrium in the absence of the wave. The force will always be such as to move the electron back toward that position, so it can be incorporated in Newton's second law of motion as follows:

$$ma \equiv m\frac{d^2s}{dt^2} = eE_0 \sin \omega t - As, \tag{20.2}$$

m being the electron mass, a being its acceleration, and A being a constant of proportionality. This is recognizable as an inhomogeneous second-order differential equation with constant coefficients. It is in fact the equation of forced simple harmonic motion.

Exercise: Show that the solution $s(t)$ to equation (20.2), satisfying the simplest initial conditions $s(0) = 0$ and $s'(0) = 0$, is given by

$$s(t) = \frac{eE_0}{A - m\omega^2}\left[\sin \omega t - \sqrt{\frac{m}{A}}\,\omega \sin\left(\sqrt{\frac{A}{m}}t\right)\right], \tag{20.3}$$

provided that $A \neq m\omega^2$. In the event that these quantities are equal—a case known as resonance, which will not be pursued here—the solution can be found directly from the original differential equation or by applying L'Hôpital's rule to the solution (20.3).

Exercise: Show that the solution for $s(t)$ when $A = m\omega^2$ is given by

$$s(t) = \frac{eE_0}{2A}\left[\sin\left(\sqrt{\frac{A}{m}}t\right) - \sqrt{\frac{A}{m}}t\cos\left(\sqrt{\frac{A}{m}}t\right)\right]. \tag{20.4}$$

Returning to equation (20.3), we note that, because of the "tight binding" of electrons in most aerosols, the term $(\sqrt{m/A})\omega \ll 1$, so the second term usually can be neglected. Then the electron *acceleration* can be approximated by the expression

$$a = -\frac{\omega^2 e E_0}{A} \sin \omega t. \qquad (20.5)$$

The amplitude of this acceleration is proportional to $\omega^2 E_0$. Recall that the electric field (and hence acceleration of the electron) is perpendicular to the direction of the incident wave. Reversing this, we anticipate that the electric field of the scattered wave will be proportional to the *perpendicular* component of the acceleration at large distances from the particle, that is, to $a \sin \theta_z$, where θ_z is the angle between the direction of the scattered light and the z-axis (see Figure 20.4), which is the direction of the incident electric field. We also expect that the total electric field is proportional to the number of electrons present in the particle, and therefore to its volume V. The *intensity* I_s of this scattered light is proportional to the square of the electric field, and allowing for the usual inverse square fall-off with distance r, consistent with energy conservation, we arrive at the proportionality relation

$$I_s^{xz} \propto \frac{\omega^4 I_0 V^2}{r^2} \sin^2 \theta_z \propto \frac{I_0 V^2}{\lambda^4 r^2} \sin^2 \theta_z \equiv \frac{I_0 V^2}{\lambda^4 r^2} \cos^2 \theta_x, \qquad (20.6)$$

where $I_0 = E_0^2$. The scattered wave makes an angle $\theta_x = \pi/2 - \theta_z$ with the forward direction (x-axis). We have therefore obtained a fundamental result for

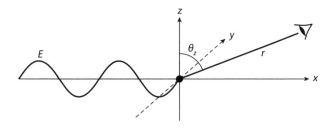

Figure 20.4. Light is scattered from a particle at the origin in a direction that makes an angle θ_z with respect to the z-axis. The observation distance is r.

Rayleigh scattering, namely that the intensity of scattered light is inversely proportional to the fourth power of the wavelength. More accurately, the probability that a photon of sunlight will be scattered from its original direction by an "air molecule" is $\propto \lambda^{-4}$. In any case, a higher proportion of blue light ($\lambda \approx 0.4\mu$) than red light ($\lambda \approx 0.7\mu$) is scattered, in fact about $(7/4)^4 \approx 9$ times more.

There are some other things to note from this formula. The scattered radiation is proportional to V^2; this results of course in a reduction of intensity in the forward direction. Note that the dependence on V^2 corresponds to one on the *sixth* power of radius R. This dependence on V can be contrasted with intensity reduction arising from *absorption* within the particle. The latter will be proportional to the number of absorbing molecules present, and therefore to V. Therefore, in relative terms the importance of scattering (as opposed to absorption) for intensity reduction in the forward direction will be greater for larger particles (still satisfying the requirement that $R/\lambda \ll 1$). Also it is clear that the intensity of scattered light depends on the angle θ_z, but remember, this is for the special case of polarized light. For scattering in the x-y plane, there is no angular dependence because of symmetry about the z-axis, and so for this case

$$I_s^{xy} \propto \frac{I_0 V^2}{\lambda^4 r^2}. \tag{20.7}$$

Sunlight is unpolarized [35] because the electric field in light from the sun vibrates in all possible directions. We can use the existing result (20.6) and generalize it to this case as follows. We define the *scattering plane* to be the plane containing both the forward and scattered directions. An unpolarized incident wave can be written as the arithmetic mean of two independent linearly polarized components, one parallel to and the other perpendicular to the scattering plane. Essentially, to do this we average the scattered intensity by replacing the term $\sin^2\theta_z$ by $(\sin^2\theta_z + \sin^2\theta_y)/2$, where θ_y is the angle the direction of observation makes with the y-axis. By including the corresponding angle with the x-axis, we note that the squares of the direction cosines sum to one, that is,

$$\cos^2\theta_x + \cos^2\theta_y + \cos^2\theta_z = 1.$$

This is equivalent to

$$\sin^2\theta_y + \sin^2\theta_z = 1 + \cos^2\theta_x.$$

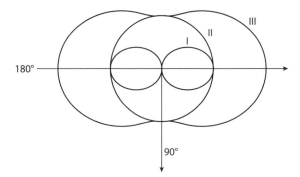

Figure 20.5. Angular variation of the polarization intensity components for the scattered light. The central curve, I, represents the component I_s^{xz} as given by equation (20.6); curve II is for the component I_s^{xy} as given by equation (20.7); and curve III is for the total intensity polarization, I_s^{T}, as given by equation (20.8). The forward direction is $\theta_x = 0°$.

Therefore in total

$$I_s^{T} \propto \frac{I_0 V^2}{\lambda^4 r^2}\left(1 + \cos^2\theta_x\right) \qquad (20.8)$$

for unpolarized sunlight. This total scattering intensity is just the sum of the two polarization intensities. This is illustrated in Figure 20.5. The perpendicular component of polarization (I_s^{xz}) is the central figure, the middle figure is the parallel component (I_s^{xy}), independent of angle because this is scattering in the x-y plane, perpendicular to the electric field vector), and the outer figure (I_s^{T}) is the sum of these two. Therefore the scattered light is most polarized (and least bright) perpendicular to the incoming light, and least polarized (and most bright) in the forward and backward directions.

The angular dependence of the polarization for unpolarized incident light is conveniently expressed in terms of the quotient

$$P = \frac{I_s^{xy} - I_s^{xz}}{I_s^{xy} + I_s^{xz}} = \frac{1 - \cos^2\theta_x}{1 + \cos^2\theta_x}. \qquad (20.9)$$

The graph of P is shown in Figure 20.6 (in Cartesian form with θ_x in radians), illustrating afresh that (for *ideal* molecular scatterers) the light is 100% polarized at 90° to the incoming radiation and completely unpolarized in the

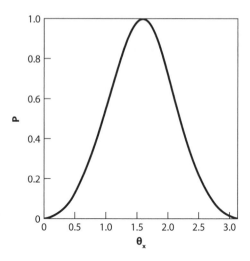

Figure 20.6. Degree of polarization for unpolarized incident light as given by equation (20.9).

forward (0°) and backward (180°) directions. (In practice this will not be the case because molecules are not perfect scatterers [36].)

We have stated that the results above for Rayleigh scattering apply when the particle size is much less than the wavelength of incident light, that is, $R/\lambda \ll 1$. But what does the phrase "much less" really mean? There is no universal definition, and as usual, it depends on the context. Several books and articles on the topic of scattering use the inequality $R < \lambda/20$, or even $R < \lambda/10$, and we shall be content with this. Since the wavelength of visible light lies in the range 400–800 nm (0.4μ–0.8μ), this means that $R < 0.04\mu$ should suffice.

For particles larger than this the processes of reflection, refraction, and diffraction cannot be neglected. It turns out that most of the light is scattered near the forward direction, but for particles whose size is comparable with the wavelengths of light, the situation is quite complex (but very interesting!) both mathematically and physically. The exact solution to this problem for spherical particles is often referred to as the Mie solution, from a 1908 paper by Gustav Mie; Debye also solved the problem independently in 1909, but the solution had already been published in 1863 by Clebsch, and again independently by Lorenz in 1890 and 1898. The rediscovery of existing results is a rather common theme in the history of science and mathematics.

The total amount of light scattered by a spherical particle is quite sensitively dependent on R/λ as already noted. This dependence can be characterized by a quantity C_s, the *scattering area coefficient*. It is defined to be the amount of light that would be intercepted by a particle (not necessarily a sphere) of cross-sectional area $C_s \pi R^2$. If, for example, the particles are very small, that is, $R/\lambda \ll 1$, the dominating mechanism is Rayleigh scattering, with its dependence on R^6 (or V^2, see equation (20.8)), then C_s will increase as $(R/\lambda)^4$ initially. For scattering by sufficiently large particles, C_s approaches two. This is the source of a fascinating paradox, *the extinction paradox*: why is C_s not equal to one? The debate involves some very sophisticated mathematical physics that will not be discussed here except to say that it involves interactions between the external and internal radiation fields of the particle. However, it should be pointed out that the paradox has not been entirely resolved if recent literature on the topic is anything to go by!

Figure 20.7 shows a graph of C_s vs. $2\pi R/\lambda$ (the ratio of circumference to wavelength) for a spherical particle with the same density as water. As noted above, C_s rises toward its first (and largest) maximum, it does so initially as the power function $(2\pi R/\lambda)^4$. This rapid increase in scattering "efficiency" reaches

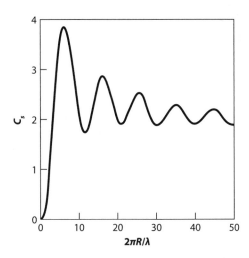

Figure 20.7. Dependence of the coefficient C_s on the ratio $2\pi R/\lambda$ for a spherical particle with refractive index equal to that of water, that is, $\approx 4/3$. (The incident light is assumed to be monochromatic for simplicity.) Redrawn from Williamson (1973).

a maximum when $R/\lambda \approx 6.3/2\pi \approx 1$. Recall that the Rayleigh scattering formula was derived assuming R/λ is small. For large enough particles the electric field within it is no longer approximately uniform, and this detracts from the cumulative effect for smaller ones, leading to a dramatic decline in C_s. The subsequent oscillatory behavior is a consequence of the particle size periodically meeting the requirements for destructive and constructive interference of the electric field within it.

Consider the graph on the left side of the first maximum, say $0 < 2\pi R/\lambda < 4\pi \approx 12.6$, and suppose that R is *fixed*. Then for longer wavelengths, corresponding to orange or red light, the scattering (as measured by C_s) is less than for shorter wavelengths, such as in blue or violet light, for that size particle (because C_s increases as R/λ increases). Now we let R be a somewhat larger (but still fixed) value corresponding to the decreasing part of the graph after the first maximum. This time the scattering efficiency is reversed: C_s decreases as λ decreases. It should be apparent that, given a suitable size distribution of particles—such as those from forest fires or volcanic eruptions—the sun or moon may appear strangely colored—blue moons and even green ones have been reported [36].

While the amount of light scattered by a particle is of theoretical interest, the amount scattered by a given mass concentration of particles is of more direct interest in gauging the effects of air pollution. For simplicity, we consider the particles to be of uniform size and density ρ. Then the mass of each particle is $m = 4\pi\rho R^3/3$. If M is the particle mass concentration (the mass of pollutant per unit volume, for example, c mg/m^3), then a cylinder of cross-sectional area 1 m^2 will contain $M/m = 3M/4\pi\rho R^3$ particles for every meter of its length. Multiplying this number by the effective particle cross-sectional area $C_s \pi R^2$ gives a relative measure of the amount f of the incident beam that is scattered, that is,

$$f = \frac{3MC_s}{4\rho R} = \frac{3M}{4\rho\lambda} \cdot \frac{C_s}{(R/\lambda)} \propto \tilde{f} = \frac{C_s}{(2\pi R/\lambda)} \qquad (20.10)$$

for a given wavelength of light. Figure 20.8 shows a graph of this ratio denoted by $\tilde{f}$; valid for any prescribed wavelength, it shows how scattering of light for a given mass concentration is most efficient when the wavelength of light is close to the particle size, that is, $\lambda \approx R$.

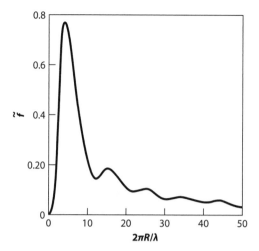

Figure 20.8. Dependence of the relative scattering coefficient $\tilde{f}$ on the ratio $2\pi R/\lambda$ for a spherical particle with refractive index $\approx 4/3$. Redrawn from Williamson (1973).

$X = x_B$: RAINBOWS AND "ROADBOWS" IN THE CITY

We are familiar with rainbows and, perhaps to a lesser extent, with atmospheric halos formed during the day when (very nearly) parallel light from the sun is "scattered" into the observer's eye by raindrops and ice crystals, respectively. The drops or crystals producing the phenomena are located on the surface of an apparent cone; in the case of the rainbow the axis extends from the observer's eye toward the so-called anti-solar point (i.e., extending to the shadow of the observer's head and beyond) as shown in Figure 20.9. In the case of, say, the common 22° ice crystal halo (Figure 20.10), the axis of the cone extends in a direction from the observer's eye to the sun.

The reader is encouraged to consult Appendix 11 if he or she needs to brush up on the basic mathematics behind the formation of these beautiful images.

But have you ever noticed a rainbow-like reflection *from a road sign* when you walk or drive by it during the day? Tiny, highly reflective spheres are used in road signs, sometimes mixed in paint, or sometimes sprayed on the sign. And sometimes, after a new sign has been erected, quantities of such

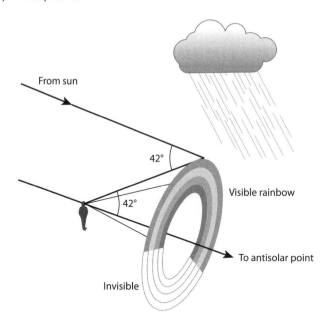

Figure 20.9. Formation of a (primary) rainbow.

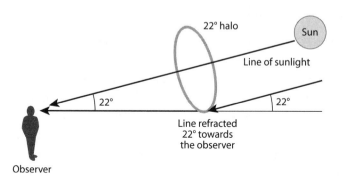

Figure 20.10. Formation of the common 22° ice crystal halo.

"microspheres" can be found on the road [37] near the sign. It is possible to get samples of these tiny spheres directly from the manufacturers, and reproduce some of the reflective phenomena associated with them. In particular, for glass spheres with refractive index $n \approx 1.51$ scattered uniformly over a dark matte plane surface, a small bright penlight provides the opportunity

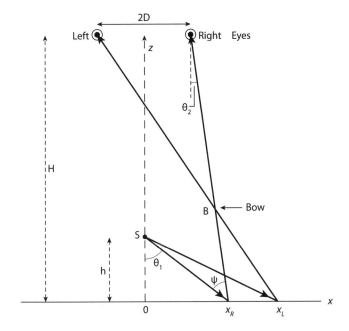

Figure 20.11. Geometry for the "microsphere" bow.

to see a beautiful near-circular bow with angular radius about 22°! (Don't confuse this with the 22° ice crystal halo; again, see Appendix 11.) This bow appears to be suspended above the plane as a result of the stereoscopic effects we now proceed to examine.

In Figure 20.11, representing the x-z plane, the (point) source of light is at the point $S(0, 0, h)$ and the observer's right (R) and left (L) eyes, aligned along the x-axis, are located at $(\pm D, 0, H)$, respectively, where $H > h$. The intersection of the bow with this plane is at the point $B(x_B, 0, z_B)$. The ray path $S x_R R$ corresponds to a deviation of $\delta = 180° - (\theta_1 - \theta_2)$ from the original direction. It is an interesting exercise to show that for a sphere of refractive index n this deviation is

$$\delta = 2 \arccos\left[\frac{1}{n^2}\left(\frac{4 - n^2}{3}\right)^{3/2}\right]. \tag{20.11}$$

For $n \approx 1.51$. $\delta \approx 2.76$ radians, or approximately $158°$.

Exercise: Use equations (A11.3) and (A.11.4) in Appendix 11 to establish the result in equation (20.11). (The angle δ here is $D(i_c)$ in equation (A11.4).)

We shall work with the angle $\psi = \theta_1 - \theta_2 \approx 22°$ in what follows. From Figure 20.8 we note that

$$\psi = \arctan\left(\frac{x_R}{b}\right) - \arctan\left(\frac{x_R - D}{H}\right). \tag{20.12}$$

Similarly, for the left eye,

$$\psi = \arctan\left(\frac{x_L}{b}\right) - \arctan\left(\frac{x_L + D}{H}\right). \tag{20.13}$$

After some rearrangement it follows that

$$T \equiv \tan\psi = \frac{(H - b)x_R + Db}{(x_R - D)x_R + Hb}. \tag{20.14}$$

The resulting quadratic equation for the location of x_R is

$$Tx_R^2 - (TD + H - b)x_R + b(TH - D) = 0. \tag{20.15}$$

All the coefficients in this equation are known, or in principle measurable, so an explicit analytic expression for x_R can be obtained. However, care is required in choosing the correct root, since x_R must be unique; it is therefore probably better to solve (20.12) directly instead. By changing the sign of D in both these equations we can obtain those for x_L also.

Using similar triangles, points (x, z) on the line Rx_R satisfy the equation

$$\frac{z}{H} = \frac{x_R - x}{x_R - D}, \tag{20.16}$$

and, changing the sign of D for x_L on the line Lx_L,

$$\frac{z}{H} = \frac{x_L - x}{x_L + D}. \tag{20.17}$$

Since these two lines intersect at the perceived bow, the point B, we can equate the two expressions and solve for $x = x_B$. Thus we find

$$\frac{x_B}{D} = \frac{x_L + x_R}{x_L - x_R + 2D}, \tag{20.18}$$

and substituting this value into either of the equations (20.16) or (20.17), we obtain

$$\frac{z_B}{H} = \frac{x_L - x_R}{x_L - x_R + 2D}. \tag{20.19}$$

Note that, from equations (20.12) and (20.13), both x_L and x_R are functions of h, H, and D.

Regarding the shape of the bow, for an observer directly over the light source it may appear to be exactly circular because of the bilateral symmetry of the human face. In the above scenario the eyes are separated by a distance $2D$ along the x-axis, but have no separation in the y-direction. Obviously, if $D \to 0$ the shape will approach that of a circle, but of course the stereoscopic effect will also disappear (and we would lose an eye!). If D were to increase, the bow would become more ellipse-like. In practice, for most observers the departure from a circular shape will be small unless the eyes are not above the light source, in which case the bow has an ellipse-like shape, and appears tilted relative to the x-y plane. (See Crawford 1988.)

Chapter 21

NIGHTTIME IN THE CITY—I

What is the difference between night and day in the city? That may appear to be a silly question, but humor me here. During the day, the sun provides the light we need, outdoors at least. Artificial lighting may be necessary inside a building, depending on the number and location (or even existence) of windows. In the country at night, the only sources of light are the moon, planets, and stars, when they are visible, barring the occasional encounter with a UFO. By contrast, most of the heavenly bodies are not seen in the city (with the possible exception of the moon and a bright planet or star). The sources of exterior light—in particular, street lamps—are artificial and, compared with the sun, relatively close by! This means that the light "rays" cannot be considered parallel, unlike those from the sun, and so the light emanating from them is divergent. This fact can give rise to some interesting modifications of certain

Figure 21.1. New York City skyline at night. Photo by Skip Moen.

daytime optical phenomena, as we shall see below. But let's see what the night brings by concentrating first on some of the less exotic but more familiar "night light" phenomena.

X = *l*: SHADOWS

As I walk home in the dark, the length of my shadow (*l*) appears to be increasing faster and faster as I walk at a constant speed *v* away from a street lamp almost right behind me. Is this in fact true? Let's see. From Figure 21.2, by similar triangles,

$$\frac{l}{h} = \frac{l+x}{L} = \frac{x}{L-h},$$

(21.1)

where *h* is my height, *x* is my distance from the base of the lamp, and *L* is the height of the lamp. Therefore

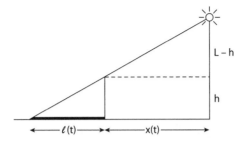

Figure 21.2. Shadow geometry for a lamp at height L.

$$l = \frac{hx}{L - h}.$$ (21.2)

Using the fact that $v = dx/dt$, the rate of increase of my shadow length is

$$\frac{dl}{dt} = \frac{hv}{L - h}$$ (21.3)

which is *constant*, so the apparent acceleration of shadow length must find its explanations in the realm of psychology and perception, not physics and mathematics . . .

Exercise: It is more frequently the case that the base of the lamp is some fixed distance y from me when I walk right past it. (I try to make sure that $y \neq 0$.) Generalize the above argument to show that

$$\frac{dl}{dt} = \frac{hvx}{(L - h)(x^2 + y^2)^{1/2}} = \frac{hv \cos \theta}{L - h},$$ (21.4)

where $\theta(t)$ is the angle the shadow makes with my direction of travel. Clearly this result approaches that of (21.3) as x increases.

Question: Will the shadow of my head move parallel to me as I walk?

X = α: DOUBLE IMAGES FROM PLATE GLASS DOORS AND WINDOWS

This is a nice little "inverse problem." Sometimes we may see a double image of the moon or a distant lamp from a window as we walk along a road at night.

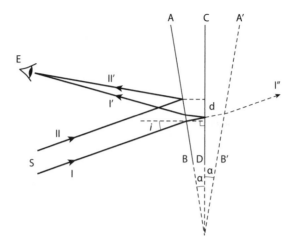

Figure 21.3. Geometry for multiple reflections in (slightly) nonparallel window surfaces for angle of incidence i (relative to the back surface **CD**). The (highly exaggerated) angle between the front and back surfaces is α radians.

Although this can occur if the window has plane parallel surfaces, it is a very tiny effect unless the surfaces are in fact not parallel. Suppose that the front and back faces of the window (*AB* and *CD* respectively in Figure 21.3) are inclined at a very small angle α, so the angle between the reflected rays I' and II' will be 2α. This is seen by calculating the deviation of a single reflected ray when the "mirror" is rotated through an angle α; using the fact that the angle of reflection is equal to the angle of incidence, it is 2α. In the (greatly exaggerated) Figure 21.3, a pair of parallel rays, I and II, from an assumed distant source are shown impinging on the back and front surfaces of the plate glass, respectively. If we imagine the back surface *CD* to be a mirror (which in effect it is), it will give a reflected image of the surface *AB* along $A'B'$, and an image of the ray I' along I''. Then the ray II'' has effectively passed through a narrow prism with apex angle 2α, and the minimum angular deflection in the ray path caused by such a prism is $2(n-1)\alpha$, which when added to the 2α above gives a total contribution of $2n\alpha$, $n \approx 1.5$ being the refractive index of the glass. Thus if we can estimate the angular distance (3α) between the two reflections, the angle between the surfaces is one third of that amount.

Exercise: The *minimum* angular deflection D in the symmetric path of a ray through a triangular prism of apex angle γ can be shown to be

$$D = 2\arcsin\left(n\sin\frac{\gamma}{2}\right) - \gamma,$$

where n is the refractive index of the prism. Show that for the plate glass window this reduces to $D = 2(n-1)\alpha$ as stated, if α is small.

How large might the angle α be and how might we estimate it? Nowadays plate glass windows are made to a very high degree of precision, but in older houses there can be some small but significant angles between the surfaces. There are probably other distortions in the thickness of the glass, but these will be ignored here.

Suppose we are close enough to estimate the lateral distance d between the images on the glass, and that the distance between the eye and the window is L. From Figure 21.4, representing the approximate geometrical relationship between L, d, and the rays I', II'', we see that $p \approx d\cos i \approx L\tan 3\alpha \approx 3\alpha L$, so that the angle between the faces is

$$\alpha \approx \frac{d\cos i}{3L}. \tag{21.5}$$

Suppose that $L = 10$ m, and $d = 1$ cm $= 10^{-2}$ m for an angle of incidence $i \approx 30°$. We now see that $\alpha \approx 3\times 10^{-4}$ radians (or about one minute of arc). Over a distance of 30 cm (about a foot) then, and on the basis of these figures, the thickness of the glass changes by about $30\sin\alpha \approx 30\alpha \approx 10^{-2}$ cm or 0.1 mm, a tiny amount, to be sure! So, if the glass surfaces are very nearly parallel (as here), it follows that the angle α is extremely tiny indeed, and the angular distance

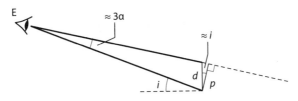

Figure 21.4. Detail for equation (21.5).

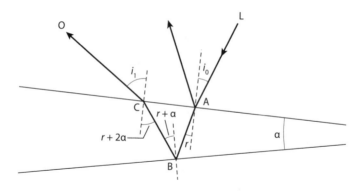

Figure 21.5. Generic ray paths for a wedge of angle α.

between the two reflections is three times "extremely tiny," that is, still very tiny! So the double images that can be observed from large distances are almost inevitably produced by nonparallel glass surfaces. If we and the light source are both close enough to parallel glass surfaces, however, then the images may then be quite well separated if the light is not at normal incidence to the window.

Let's pursue this a little farther. From Figure 21.5 we see that the ray $LABCO$ exits the upper glass surface at an angle i_1 to the normal direction, where $\sin i_1 = n \sin (r + 2\alpha)$. Also, $\sin i_0 = n \sin r$. Using the fact that α is a very small angle, we may write that $\sin i_1 \approx n \sin r + 2n\alpha \cos r$. The difference between the angles i_0 and i_1 is small, so we can use differential notation to write

$$\sin i_1 - \sin i_0 \approx \delta(\sin i) \approx 2\alpha n \cos r.$$

Therefore

$$\delta i \approx \frac{2\alpha n \cos r}{\cos i} = \frac{2(n^2 - \sin^2 i)^{1/2}}{\cos i} \cdot \alpha \equiv k(i)\alpha. \tag{21.6}$$

Note that $k(0) = 2n$. That is, for normal incidence, the angular separation of the images is approximately $2n\alpha = 3\alpha$ for $n = 1.5$ (a generic value for the refractive index of glass). This is the result obtained earlier in this section. A graph of the function $k(i)$ is shown in Figure 21.5. Clearly, this indicates that δi, the approximate angular separation of the images, becomes larger and larger the closer the ray's angle of incidence is to the plane of the surface.

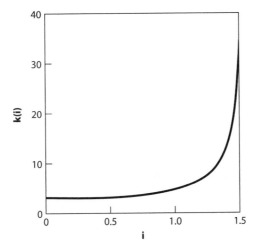

Figure 21.6. The function $k(i)$ in equation (21.6).

There are two additional factors that can modify the approximation (21.6), one being due to the divergence of the rays from a light source which is in practice *not* infinitely far away (and neither is the observer!). Suppose that the light source L is a distance R_L from the glass, the distance of the observer O is R_O, and the distance between these two points on the glass is d. Using simple geometrical optics for a single ray from the light source L to the observer's eye at O (via the point P on the glass surface) we can show that (Figure 21.7)

$$f(R_L, R_O) = \frac{R_L}{R_L + R_O} = \frac{d - x}{d}. \tag{21.7}$$

The factor f is basically a measure of the angle of incidence for an individual surface-reflected ray as determined by the relative positions of source and observer (and is independent of the earlier geometric argument leading to equation (21.6)). It approaches one as the distance of the light from the glass (relative to the observer) becomes larger. Conversely, as this relative distance becomes smaller, f decreases toward zero. The factors k and f in equations (21.6) and (21.7), respectively, will tend to counteract one another as the angle of incidence approaches 90°.

Another geometric factor ($\cos \phi$) comes from when the plane of incidence (containing the incident and reflected rays) is at an angle ϕ to the plane

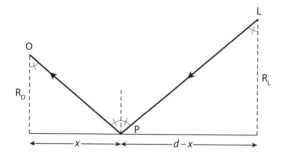

Figure 21.7. The basic observer (**O**)-light source (**L**) geometry, neglecting the effects of wedge refraction for the additional factor **f** in equation (21.7).

containing the wedge angle of the glass (i.e., the plane is rotated by an angle ϕ about the normal to the glass pane). This factor will reduce the angular distance between the two images unless $\phi = 90°$. Taking all these factors into consideration, we have the more general approximation for δi, namely

$$\delta i \approx \left(\frac{R_L}{R_L + R_O} \right) k(i) \alpha \cos \phi. \tag{21.8}$$

$X = r(t)$: RAIN "SPARKS"

After it has been raining awhile and there are several large puddles on the road, as you walk you may notice that the light's reflection is surrounded by momentary "sparks" emanating from where the raindrops hit the surface. They appear to point radially outward. In his classic book *Light and Colour in the Open Air*, Marcel Minnaert [38] points out that "the explanation is simple." Let's see. When a drop hits the surface of the water it generates a set of concentric circular wave patterns (see Chapter 25 for more details about this). As the wave with center C expands (see Figure 21.8), a sequence of reflections occurs in an almost continuous fashion along the line CM (C_1 and C_2 are intermediate points of reflection). As the wave expands outward, so does the point of reflection, giving the appearance of rapidly moving sparks of light.

A related reflection phenomenon occurs when a street lamp is observed through the crown of a tree (Figure 21.9a). The light is reflected into the observer's eyes by twigs and leaves, and the effect is enhanced if the tree is wet

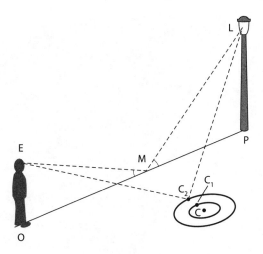

Figure 21.8. "Rain spark" geometry.

after rain, or even frost-covered. As indicated in Figure 21.9a, all the branches and twigs in a given plane may contribute reflections to the observer, but those parallel to the line EL will be greatly foreshortened compared with those perpendicular to that line. Those between these orientations will be foreshortened proportionately less. As such, the combined effect will be to create a "halo" effect—a set of approximately concentric circles (or arcs of such circles).

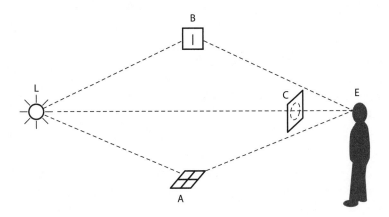

Figure 21.9a. "Rain circle" geometry when a source of light shines through wet leaves and twigs.

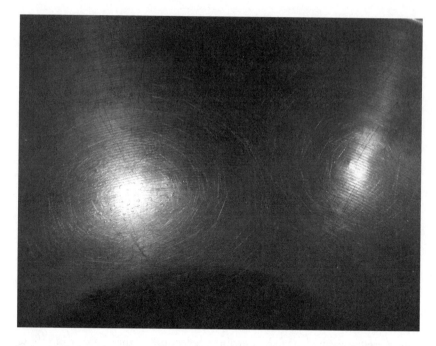

Figure 21.9b. Scratches in a stainless steel bowl exhibit circular and elliptical patterns similar to the "rain circle" phenomenon in Figure 21.9a. In this picture reflections from two kitchen lights produce the effect.

A very similar phenomenon occurs when you look at the sun through a train or airplane window. The many fine scratches give rise to a similar effect, and for very similar reasons; we only notice those scratches that are perpendicular to the plane of incidence of the light rays. As an example of the same type of thing arising in "kitchen optics," Figure 21.9b shows a similar "rain circle" effect from scratches in a stainless steel mixing bowl. The "circles" are really more elliptical in shape because of the curvature of the bowl.

X = *R*: REFLECTIONS IN THE RAIN

On a rainy night, light from street lamps reflected by power lines and other cables may be noticed by the observant pedestrian. Let's assume that you are that person. In Figure 21.10, what is the relationship between your position (at O say), that of the light and that of the reflection P, if any? Since any three noncollinear points define a plane (if we are permitted to assume that your

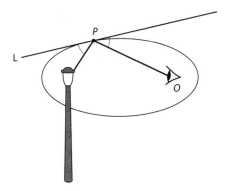

Figure 21.10. The reflection **P** of the lamp in a power line; the lamp and observer"s eye at **O** are at the foci of the ellipse.

eyes, the light, and its reflection are points here!), and you see the reflection, it follows that *P* lies on an ellipse, with you and the light source as foci. Why is this so? One fundamental property of an ellipse is that the lines joining the foci to any point on the ellipse make equal angles with the tangent line at that point (see Figures 21.11a and 21.11b), and hence with the normal line at that point. This is the law of reflection expressed as a property of the ellipse.

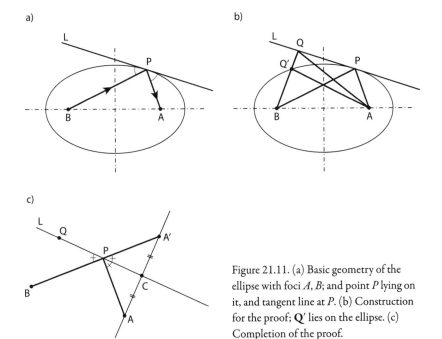

Figure 21.11. (a) Basic geometry of the ellipse with foci *A*, *B*; and point *P* lying on it, and tangent line at *P*. (b) Construction for the proof; **Q′** lies on the ellipse. (c) Completion of the proof.

This is proved geometrically as follows. P is a point on the ellipse, with tangent line L, along which the point Q can move. When Q is coincident with P, the distance $AQ + BQ$ is a minimum by virtue of the property that $AQ' + Q'B =$ constant. Now let A' be the reflection of the point A in the line L, so that $AC = CA'$. Then $AQ + QB = A'Q + QB$ which is a minimum when $Q = P$, and BPA is a straight line. This being the case, angles APC and CPA' are equal, and hence so are the angles APC and QPB. The reflective property of the ellipse is established.

Exercise: Prove this result algebraically (you'll also need a modicum of calculus).

Chapter 22

NIGHTTIME IN THE CITY—II

In a city at night, owing to artificial light sources, we may witness substantial modifications to light pillars, ice crystal halos, and rainbows when the weather conditions are right for producing them. The sources of light are much nearer than the sun (or moon) and thus the light from them is divergent, not parallel. This can give rise to some quite amazing three dimensional "surfaces" that differ significantly from their daytime (parallel light) counterparts. In order to appreciate this, some geometry associated with surfaces of revolution will be required. But before getting into that, let's ask a much more basic question as an appetizer to the main course, a question that is surely on the minds of everyone in these dark times:

Question: How many light bulbs are there in Times Square? [6]

This is a question that has been radically changed by technology because there are so many types of "light bulbs." Most of the illumination in Times Square comes from giant electronic billboards rather than from traditional light bulbs. These billboards emit their own light. In fact, each pixel of these billboards does so and could be counted as a separate "light bulb." To answer the question, then, let's estimate the number of these pixels. Each electronic billboard has about the resolution of a computer monitor or about 1000×1000 pixels = 1 million pixels. There are definitely more than one and fewer than 1000 billboards, so again, let's use the geometric mean of 30. This gives about 30 million separate illuminated pixels.

Looking at a picture of Times Square, one can't see that many traditional light bulbs. There are a few scrolling message signs on marquees. They are about 1 letter tall and 50 or so letters long. Each letter will have at least $6 \times 8 \approx 50$ light bulbs, so let's round up to 100. This gives 5000 light bulbs per scrolling message. There are five or ten of those, so that gives another 40,000 light bulbs. The static illuminated signs (traditional neon lights or plastic signs back-lit by fluorescent bulbs) will probably add another few thousand, but that's just a rounding error.

Now we can start thinking about the effects of light on ice crystals in the night air—brrr!

$X = \alpha_{max} - \alpha_{min}$: LIGHT PILLARS FROM STREET LIGHTS

As plate-like ice crystals fall they can reflect and refract light. In the daytime *sun pillars* are produced by reflection from the faces of tilted crystals in the vicinity of the rising or setting sun. Specifically, the more common upper pillars result from the reflection of sunlight downward from the lower faces of such crystals, and lower pillars arise from sunlight being reflected upward from the upper faces of tilted crystals. Upper pillars can become brighter as the sun descends below the horizon, and can appear to extend very high up in the sky if the upper crystals have larger tilts than the lower ones (see Adam 2011 and references therein). Artificial sources of light can also produce light pillars as shown in Figure 22.1. The angular extent of the pillar is the maximum range of angles (α) subtended at the observer's eye by light reflected from the crystals.

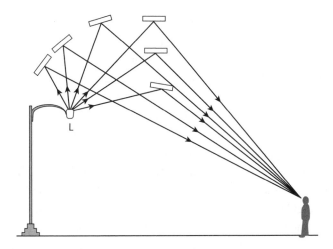

Figure 22.1. How light pillars are formed from ice crystals in the vicinity of a lamp. Each "ray" entering the observer's eye makes a different angle α with the horizontal direction.

This is determined by their distribution in the air between the observer and the light source (though some beyond the source may also contribute). The angular width of the pillar is similarly defined.

And *now* for some of the more esoteric artificial light phenomena, as promised earlier.

Question: What is a cone during the day and a cigar or an apple at night?

Answer: The locus of directions along which one sees an ice crystal halo or a rainbow.

Let's unwrap this conundrum a little bit. When you look at a small portion of a rainbow, you are looking in a particular direction. Sunlight is scattered by raindrops from this direction into your eyes. For the primary (brighter) bow this direction is at about 42 degrees from the imaginary line joining the sun to the shadow of your head (the antisolar point). Scattered sunlight coming from all other directions at 42 degrees to this line will also reach your eye, resulting in the familiar circular arc we see as a rainbow. Thus we are looking along the *surface of a cone* with half-angle approximately 42 degrees. The same principle applies to the smaller 22-degree ice crystal halo commonly seen around the sun, or the moon at night; again, we are looking along the surface of a (smaller) cone when we observe this halo. We can think of the rainbow and halo arcs in another way: as the intersection of the imaginary cone with an imaginary plane perpendicular to its axis. And no matter how near or far away the raindrops or ice crystals may be, they give rise to the same circular arcs, because we are looking in *certain directions*; furthermore, rainbows and halos are not *objects*, they are images! Put another way, you cannot "back up" to get more of a rainbow in your camera viewfinder. Of course, if you change your lens for a wide angle one, you can achieve that result! Note that the rainbow, an arc of a circle of angular radius 42° around the antisolar point, is the same as a circle of 138° around the sun. Thus a cone with 'half-angle' 138° opens up to the opposite direction; this establishes the notion that bows and halos are intimately connected in a geometrical sense as well as an optical one. This will be a valuable unifying concept in the next section.

And now, to geometry!

X = r_{max}: "CIGARS AND APPLES" IN THE CITY

An important theorem in plane geometry states that the angle subtended by a chord (CE) at the center of a circle (2α) is twice that subtended at any point (F) on the circumference. This is easily established from Figure 22.2 by adding the radius OF to form three isosceles triangles, and noting that the sum of the angles around O must be 360°. Now let C represent a light source, and E

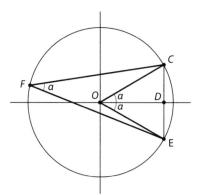

Figure 22.2. Angle subtended by a chord (**CE**) at the center of a circle (2α) is twice that subtended at any point (**F**) on the circumference.

represent the observer's eye. If the volume of space surrounding the observer and source contains either raindrops or hexagonal ice crystals (and is sufficiently large), then for raindrops, $\alpha \approx 42°$ and the arc *CFE* is the locus of all raindrops in that plane scattering light into the observer's eye. But by symmetry this is also true for all rotations of that arc around the line *CE*, forming, as it turns out, an apple-shaped surface. For reasons that will become obvious below, this surface is known as "Minnaert's cigar."

Reference has already been made to Marcel Minnaert, the author of a famous and widely quoted book *Light and Colour in the Open Air* [38]. In that book (pp. 206–207) he made the following observation:

> One very cold evening (17° F) beautiful halo phenomena could be seen in the steam from a train in the railway station. Near one of the lamps, where the cloud of steam was blown in every direction, a cigar-shaped surface of light could be seen, having one end near the eye, the other near the lamp; all little crystals traversing this surface were lit up, but the space inside was quite dark; the cone tangential to the surface had an angle at the vertex of about 44°. It is at once clear that the cigar-shaped surface is simply the locus of all those points P such that the sum of the angles subtended by EP and PL at L and E respectively is 22°.

Minnaert then goes on to note that the remarkable three-dimensional nature of this observation is only possible because (i) the light source is so near the eyes, and (ii) of the stereoscopic effects associated with both eyes viewing the crystals (see Figure 22.3).

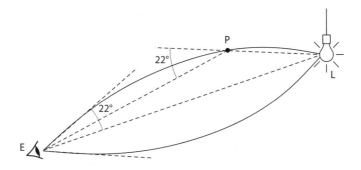

Figure 22.3. Sketch of "Minnaert's cigar" (Redrawn from Minnaert 1954).

Figure 22.4. Minnaert"s cigar for the 22°
halo, illustrating the three-dimensional form
(Redrawn from Mattsson et al. 2000).

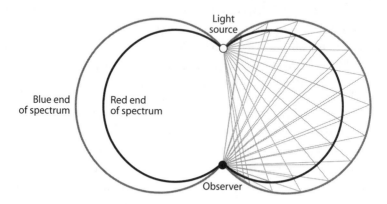

Figure 22.5. Minnaert's "apple-shaped cigar" for the 22° halo, generated by rotating the larger arc around the chord **CE** in Figure 22.2 (or chord **LO** in Figure 22.6). The light source is at the top of the "core" and the observer"s eye is at the bottom. It is drawn for both the red and blue ends of the visible spectrum. Redrawn from a diagram by Christian Fenn (see [39]).

Figure 22.4 is a perspective rendering of the Minnaert 22° cigar, a spindle–shaped surface, truncated here to emphasize its three-dimensional nature. And the apple-shaped surface in Figure 22.5 is also referred to as Minnaert's cigar (even though it looks like no cigar I've ever seen!). It arises from rotating the smaller arc CE about the chord in Figure 22.2, and the apple surface arises from rotating the larger arc CFE about the same chord.

In fact there are several other intermediate surface shapes that can arise from similar considerations. In addition to the "classical" 22° cigar halo surface (with supplementary angle 158°), there is a less common 46° one (with supplementary angle 134°). Furthermore, as also noted in Appendix 11, there is a secondary rainbow which can appear at an angle of 51° from the anti-solar direction (or as a circle of angular radius 129° around the sun). When rotated about the corresponding chord LO (see Figure 22.6, where the segments LO

correspond to the chord CE in Figure 22.2), the appropriate "cigar surfaces" are generated (Figures 22.4 and 22.7). In each case, the vertex angles marked in Figure 22.6 are the angles measured from the anti-solar direction. The scattering angle—the angle through which light from the source has been deviated—is the supplement of these, thus 22° and 46° for the halo surfaces, 138° and 129° for the rainbow surfaces.

An enlarged version of the small 22° "cigar" surface of Figure 22.6 is shown in Figure 22.7; as in Figure 22.4 it can be interpreted as a surface of revolution.

The detailed geometry of the situation is governed by Figure 22.8, where the observer's eye is at B and the light source is at D. We see from triangle BCD that $r_c + r_d = r_{max}$. From this it follows that

$$\tan r_d = \frac{d}{1-d} \tan(r_{max} - r_d). \tag{22.1}$$

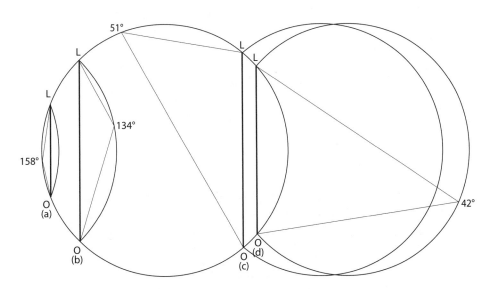

Figure 22.6. Generalization of Minnaert"s cigar for (a) the 22° halo, (b) the 46° halo, (c) the secondary rainbow, and (d) the primary bow. The light source is at **L** and the observer"s eye is at **O**. The angles stated are measured from the "anti-source" point (as in the daylight anti-solar point). Thus the supplement of the 22° halo angle is 158°, etc. (Based on Figure 1 in Mattsson and Barring, 2001).

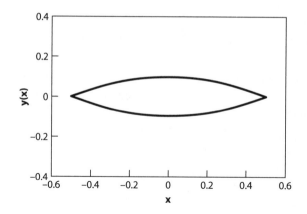

Figure 22.7. Profile of Minnaert's cigar as defined by equation (22.5).

First, we do a little geometry. The unit of length is taken as the distance BD, that is, $BD = 1$. This is the length of the "cigar." The radius of the circle is $R = BD/(2 \sin r_{max}) = (2 \sin r_{max})^{-1}$; the distance $d_c = BC$ between the crystals scattering light into the observer's eye and his eye at B is

$$d_c = (1 - d) \sec r_d. \tag{22.2}$$

The quotient q (or "aspect ratio") of the cigar's short and long axes is

$$q = \frac{R(1 - \cos r_{max})}{R \sin r_{max}} = \tan\left(\frac{r_{max}}{2}\right). \tag{22.3}$$

Let's now find the equation of this cigar-shaped halo. The center of the circle, O', in Figure 22.8 is located at the point $(0, -R \cos r_{max})$ or $(0, -(\cot r_{max})/2)$. The equation of the circle is therefore

$$x^2 + \left(y + \frac{1}{2} \cot r_{max}\right)^2 = \frac{\csc^2 r_{max}}{4}. \tag{22.4}$$

For the 22° halo, this angle is just r_{max}, so equation (22.4) can be rewritten as approximately

$$x^2 + (y + 1.238)^2 = 1.782. \tag{22.5}$$

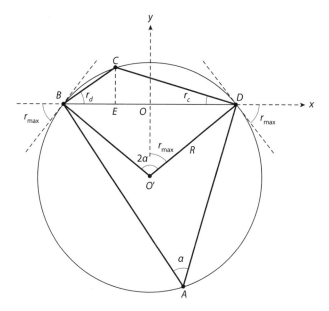

Figure 22.8. Relationship between the angular radius of a halo (r_d) with its center at a distance d from the light source. The length of the "cigar" is distance $BD = 1$, $ED = d$, angle $BCD = 180° − \alpha$, $\mathbf{O'}$ is the center of the circle and $\mathbf{O}$ is the coordinate origin. Note from triangle $\mathbf{O'OD}$ that $\alpha = r_{max}$.

This is the equation of the profile illustrated in Figure 22.7, noted earlier in "solid" form in Figure 22.4.

As noted above, Figure 22.6 illustrates a cross section of the complementary cigar shape for other situations with much smaller "scattering angles" relative to the anti-source point (as opposed to the anti-solar point during the daytime). This includes both primary and secondary divergent-light rainbows. Each apple-shaped rainbow surface (c) and (d) can be regarded more accurately as a degenerate form of a torus, missing the central hole because the rotating circle intersects itself. This is well illustrated in the next chapter (Figure 23.10).

Complicated as these surfaces may seem, there is a very interesting two--dimensional subset of them that may be seen when the ground is covered with raindrops or dew (for "dewbows") and frost or other ice crystals, such as dia-mond dust (ground-level cloud composed of tiny ice crystals; see Appendix

11). Recall that a corresponding daytime phenomenon—"roadbows" (glass bead bows)—has been mentioned in chapter 20. As there, we consider first the simpler case of (nearly) parallel light from the sun (or occasionally the moon for "moonbows"). As noted above and in Appendix 11, sunlight scattered from raindrops can produce rainbows, and the scattered light appears to lie on the surface of a cone of semi-angle 42° with vertex at the observer's eye. Now if the scattering droplets lie on the ground (as with dew), then that cone is effectively intercepted by a horizontal plane—the flat ground! If the solar altitude exceeds 42° the observer will see an elliptical "rainbow"; if it is exactly 42° the shape will be parabolic, but for an altitude less than 42° a hyperbolic rainbow or dewbow arc will be seen. This follows directly from the geometric definition of the conic sections. These same effects can be witnessed when flying above a cloud deck, by the way, for which the rainbow becomes a "cloudbow."

$X = (d/2)\csc\alpha$: DEWBOWS IN THE CITY

Now we will think about rainbows produced by sources of divergent light and scattered by droplets *on the ground*, that is, by dew. It is evident from Figures 22.2 and 22.6 that the locus of scattering drops will be the major arc of a circle (such as *CFE* in Figure 22.2) in the plane containing the observer's eye and the source of light. The chord *CE* will generally not be horizontal unless the lamp and eyes are at the same level. Therefore the intersection of this circular arc with the ground will be another circular arc of larger radius. If the distance *CE* between the light source and the observer's eye is d, then the locus of all points subtending an angle $\alpha < \pi/2$ is the major arc of a circle of radius $r = (d/2)\csc\alpha$. For generic primary and secondary rainbow angles of 42° and 51° these are approximately $0.75d$ and $0.64d$ respectively. Interestingly, therefore, in divergent light the radius of the secondary bow arc is *smaller* than that for the primary bow.

This is illustrated in Figure 22.9 for an observer at O and a light source at L. When a nonvertical chord LO lies completely above the ground, and the circular arc is rotated around this chord, there is a point at which the arc is tangential to the ground, and thereafter the intersection with the ground splits into two points as illustrated in Figure 22.9. These two points of intersection move apart and then together, eventually tracing out the kidney-shaped curves illustrated in figures (a)–(d). The traces (e) and (f) correspond to the observer being directly below the light source, and above ground. In (e) the observer is

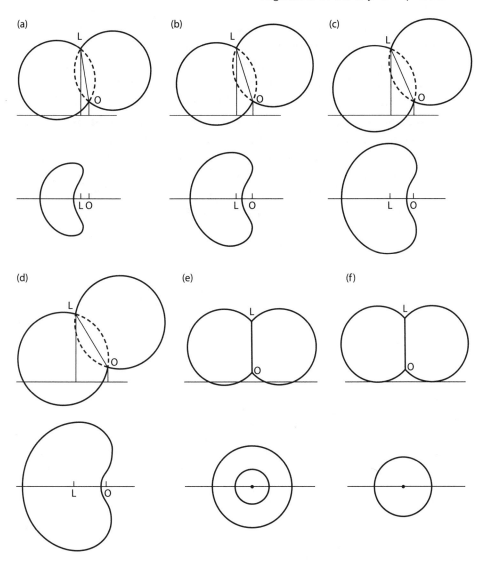

Figure 22.9. Intersections of various "Minnaert cigar rainbow surfaces" with the ground for different configurations of observer (**O**) and light source (**L**). Redrawn from Mattsson (1998).

low enough that the Minnaert "apple" surface traces out two concentric circular intersections; in (f) the surface is tangential to the ground because the observer is slightly higher above ground than in case (e). According to a report by Mattsson (1998), bows of the form (a)–(d) have been reported, but those like (e) and (f) have not. This is a challenge for the interested and observant reader!

Regarding the corresponding "ground *halos*" produced by frost or diamond dust, I have been informed by Alexander Haußmann that as young scientists, he and his friend Richard Löwenherz observed that in frost, only the 22° was frequently seen, whereas in diamond dust both the 22° and 46° halos were often present. They concluded that frost crystal faces were generally not well enough developed to ensure the consistent presence of the larger halo.

Chapter 23

LIGHTHOUSES IN THE CITY?

Lighthouses in a town or city, you ask? Certainly; there are over a thousand in the United States (though many of them are no longer in use), and Michigan has the most lights of any state, with more than 150 past and present lights. A state-by-state listing of all U.S. lighthouses may readily be found online, as well as listings for lighthouses in Europe and elsewhere. Many of these are close to or within city boundaries. London's only lighthouse (no longer functioning as such) is located at Trinity Quay Wharf in London's docklands. By contrast, New York City has several. According to one online account [40]

> When we think of lighthouses, we often conjure up images of majestic white towers perched on rocky outcroppings dotting the New

England coastline, or a stolid monolith announcing your arrival to the Outer Banks. But lighthouses in New York City? A city of soaring skyscrapers and hard concrete? What business do these anachronistic buildings of olden days have in New York, a city forever living on the cutting edge? Well, the truth is lighthouses do exist in New York City, and they have rightfully earned their place here, having helped build up this great city into what it has become.

Oh, by the way. If you still find the idea of lighthouses in a city too much of a stretch, consider *searchlights* instead. Once used primarily for military purposes, they now are widely used for advertising, fairs, festivals, and many other public events. Most of the features discussed below also apply to searchlight beams.

$X = \theta$: RAINBOWS IN LIGHTHOUSE BEAMS

Again, we can have some fun examining the "geometry of light" in this rather unusual context, and as in previous sections, we find some quite surprising results (see Figure 23.1). As a lighthouse (or searchlight) beam sweeps a

Figure 23.1. Primary (and faint secondary) bow in the beam of the Westerhever Lighthouse in Nordfriesland (Germany). Photo by Achim Christopher.

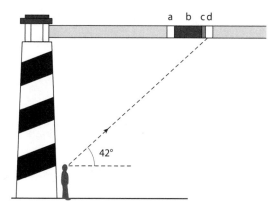

Figure 23.2. Rainbow "slices" seen in a lighthouse beam. The primary bow (or bright section) is at **d**, and if color is seen at all, the point **c** will be tinged with red. Alexander"s dark band is the segment at **b**, and the fainter secondary bow, if visible, will be at point **a**. (Redrawn from Floor (1982).

rain-filled sky, a primary and often a secondary rainbow slice may often be noticed. As with a rainbow during the day, they are separated by a dark band (see Figure 23.2). Naturally, only those drops in the beam will scatter light into the observer's eye. Subtle differences occur between the two types of beam, however, because a searchlight is close to the ground with a beam at an angle to the vertical direction, whereas the lighthouse is elevated relative to the observer, and has a horizontal beam. In each case, as the beam rotates about a vertical axis, the bows appear to slide up and down the beam.

We will use Figure 23.3 to explain this phenomenon for a lighthouse. Its beam, parallel to the ground, comes from a lamp L at height h above the ground. The observer at O is a distance d from the base of the lighthouse, standing directly under the beam at a particular instant of time. It is raining! The point O' is a distance h vertically above O. From what we already know about rainbow formation, we know that if the angle $\theta \approx 42°$, a primary bow (a better word might be slice) will be seen at B_p on the beam; similarly, if $\theta \approx 51°$, a secondary bow will be located at B_S, though it will be fainter (and may not be visible at all). Certainly the beam will by contrast be brighter there than for the angular range $42° < \theta < 51°$, corresponding to the dark region between bows (Alexander's dark band). This may be more readily seen in a lighthouse beam than in sunlight. The reddish edges of the bow define the

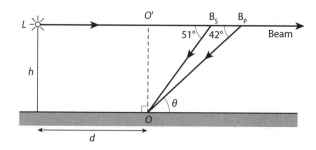

Figure 23.3. Geometry for rainbow formation in a stationary lighthouse beam.

ends of the dark band, though the colors may well be less evident because the eye is less sensitive to color at the (relatively) low light intensities present in a lighthouse beam.

When rainbows are produced by sunlight (or moonlight) the secondary bow appears higher in the sky than the primary. As such, it is often perceived to be farther away than the primary bow, but this is erroneous. While raindrops contributing to the secondary bow may or may not be farther away than those for the primary, the case for the lighthouse bows is definite and clear-cut: while the secondary bow appears higher in the sky, the drops giving rise to it *are* nearer the observer. This can be seen from Figure 23.3. The drops at B_p are a distance $OB_p = h \csc 42° \approx 1.5h$ from the observer; those at B_S are at a distance $OB_S = h \csc 51° \approx 1.3h$. But this is for the case of a beam that is stationary with respect to the observer, so unless he or she runs in circles to stay directly under the beam, we need to consider a more general case!

$X = R(\phi)$: THE THREE-DIMENSIONAL CASE

As the beam rotates, the bows will appear to slide back and forth along the beam to maintain the rainbow angles of $42°$ and $51°$, respectively. Let's examine the geometry for an observer at O positioned outside the plane containing the lighthouse and its beam, as shown in Figure 23.4. For simplicity we consider that the beam in direction LR does not diverge significantly, and thus retains its intensity over its visible length. R is the location of a patch of rain scattering the light back toward the eye of the observer. O' is a point directly

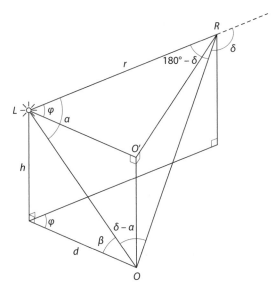

Figure 23.4. Geometry for "sliding rainbow" slices.

above the observer at the same height as the light source. The beam is at an angle ϕ from the line joining O to the base of the lighthouse. δ is the angle through which the light is deviated at R to be seen as a rainbow "slice" at O. For a primary bow, $\delta \approx 138°$. Finally, angle $O\hat{L}R = \alpha$.

From Figure 23.4, applying the law of sines to triangle LOR, we find that

$$\frac{r}{h} = \left(1 + \frac{d^2}{h^2}\right)^{1/2} \frac{\sin(\delta - \alpha)}{\sin(180 - \delta)}. \tag{23.1}$$

Application of the rule of cosines to triangles LOR and $LO'R$ respectively yield

$$(OR)^2 = r^2 + (h^2 + d^2) - 2r(h^2 + d^2)^{1/2}\cos\alpha, \text{ and}$$
$$(O'R)^2 = r^2 + d^2 - 2rd\cos\phi.$$

Noting that triangle $OO'R$ is a right triangle, it follows from these equations that

$$\cos\alpha = \left(1 + \frac{d^2}{h^2}\right)^{-1/2} \frac{d}{h}\cos\phi. \tag{23.2}$$

Now we are in a position to express the distance R of the scattering drops, relative to the height of the light source (so $R = r/h$) in terms of the relative distance of the observer, $D = d/h$, and the angles ϕ and δ after some algebra as

$$R(\phi; D, \delta) = D \cos \phi - (1 + D^2 \sin^2 \phi)^{1/2} \cot \delta. \qquad (23.3)$$

Note both from the equation and the polar plots in Figure 23.5 that R is symmetric about the polar axis, $\phi = 0$, and that if $\phi = 180°$, then $R = -(D + \cot \delta)$. Also, if $D = 0$, the plot reduces to a circle of radius $R = -\cot \delta \approx 1.11$).

What exactly do these polar graphs tell us? That's a good question. They show the distance R of the rainbow (or more accurately, that of the corresponding raindrops) in units of h along the beam as a function of beam orientation ϕ for given values of D. This is the distance (also in units of h) of the observer from the base of the lighthouse. Simply put, the graphs show the points (in the horizontal plane through L) where the observer sees the primary bow. (A similar set of graphs can be drawn for the secondary bow.) To interpret this, note that the observer is on the line $\phi = 0°$ (the positive x-axis) at a distance D from the origin at L. For the case of $D = 0.5$ the rainbow lies on a slightly squashed circle as the beam rotates; for $D = 1.11$ the rainbow moves along the beam toward L and eventually coalesces with the light source when $\phi = 180°$. The case for $D > 1.11$, (specifically $D = 2$ here) is interesting. The dotted loop represents the position of the rainbow if there were *two beams* 180° apart. When $\phi = 146°$ the rainbow would emerge from L and move outward along the "backward" beam until $\phi = 180°$ and then move back toward L and continue in the counterclockwise path as shown. When there is only one beam, the rainbow would rapidly approach L (and "disappear" for a short time by remaining there for the duration of the now "virtual" loop) before rapidly reemerging from L and proceeding in the larger loop.

But where did the angle of 146° come from? It is calculated by setting $R = 0$ in equation (23.3) and solving the resulting expression for ϕ (given a value for D), to arrive at

$$\sin^2 \phi = \left(1 - \frac{\cot^2 \delta}{D^2}\right) \sin^2 \delta. \qquad (23.4)$$

Clearly, $D > |\cot \delta|$ if ϕ is to be a real number. For $\delta = 138°$ and $D = 2$ we find that $\sin \phi \approx \pm 0.557$. Since $R = 0$ in the second quadrant, we obtain $\phi \approx 2.55$ radians, or approximately 146°. The limiting case of no loop occurs for

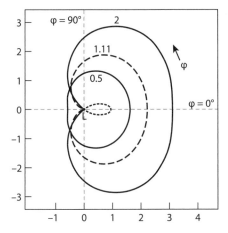

Figure 23.5. Polar diagram for the position of the bow as seen by the observer, based on equation (23.3). In decreasing order of size the graphs correspond respectively to $D = 2$, 1.11, and 0.5. In each case the "rainbow angle" is $\delta = 138°$. **L** is at the center of any circle $R =$ constant, and the observer is located on the horizontal axis at relative distance **D** from the base of the lighthouse.

$\phi = 180°$, and from equation (23.4) this corresponds to $D = |\cot \delta| \approx 1.11$. This limiting case is also shown in Figure 23.5.

Exercise: Derive equation (23.4).

$X = \theta$: HALOS IN LIGHTHOUSE BEAMS

To see halos in such beams, there must be ice crystals present and we must turn around to face the lighthouse. Strictly speaking, as with the rainbows, these are *slices* of halos because of the narrowness of the beam. And as with solar and lunar halos, it may be necessary to block out the source of light in order to see a halo clearly. The 22° halo in particular appears to be darker inside the "rim", so the beam is likely to appear darker nearer to the lighthouse, with a brighter spot corresponding to the halo slice. The larger halo will also be associated with a bright spot on the beam. As can be seen from Figure 23.6, the observer O should be at a distance of at least $h \cot \theta$ from the base of the lighthouse for halos to be visible. The angles for the 22° and 46° halos are (not surprisingly) $\theta = 22°$ and 46°, so $d > 2.48h$ and $d > 0.97h$, respectively (and approximately).

To compute the corresponding polar plots for the halos, we set $\delta = 22°$ and 46° respectively in equation (23.3). These are shown in Figure 23.7 for $D = 2$. Note that, by contrast with the rainbow slices, the halos are visible for only a short period as the beam rotates over the observer.

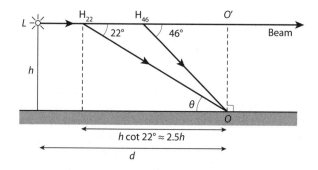

Figure 23.6. Geometry for halo formation in a stationary lighthouse beam.

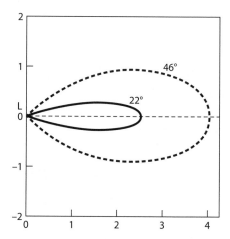

Figure 23.7. Polar diagram position of the halos as seen by the observer **O** at a distance of $D = 5$ units from the base of the lighthouse, using equation (23.3). In decreasing order of size the graphs correspond to $\delta = 46°$ and 22°.

$X = \theta$: RAINBOWS IN SEARCHLIGHT BEAMS

How do things change for the observer when looking at a searchlight beam? The situation is rather different because the searchlight will be close to the ground and pointing upward (unless we're thinking about Alcatraz in its halcyon days), and therefore the beam will not be parallel to the ground. The notation in this case will differ a little from that in Figure 23.3; the angle OSR between the observer-searchlight line and the beam will be denoted by μ (see Figure 23.8). The distance from the searchlight to the rainbow slice is r. The

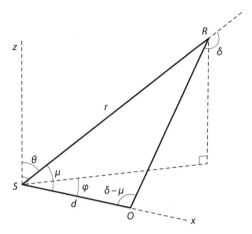

Figure 23.8. Geometry for rainbow "slice" **R** in a searchlight beam. The searchlight is at **S** and the observer is at **O**.

angle between the beam and the vertical direction is θ, and the angles ϕ and δ are as defined in the previous section. It follows that the angle $SOR = \delta - \mu$. Applying the law of sines to the triangle SOR we have that

$$\frac{d}{\sin(180° - \delta)} = \frac{r}{\sin(\delta - \mu)}. \tag{23.5}$$

This can be rearranged as

$$\frac{r}{d} = \cos\mu - \sin\mu \cot\delta. \tag{23.6}$$

From Figure 23.8 it is apparent that

$$r\sin\theta\cos\phi = d = r\cos\mu. \tag{23.7}$$

By substituting for μ from equation (23.7) into (23.6) we obtain the following expression for the relative distance R of the rainbow slice along the beam, where $R = r/d$:

$$R = \sin\theta\cos\phi - (1 - \sin^2\theta\cos^2\phi)^{1/2}\cot\delta, \tag{23.8}$$

noting again that $\delta \approx 138°$ for the primary bow. The polar graphs for R are qualitatively similar to those for the lighthouse beam in Figure 23.4, and will not be reproduced here. Again, "virtual" loops can appear in the plots if the angle θ is too large, corresponding the bow disappearing for a small range of the ϕ on the far side of the beam (see Harsch and Walker (1975)). This does not occur for $\theta < 48°$ approximately, as may be verified by setting $R = 0$ and $\phi = 180°$ in equation (23.8). The analysis for a secondary rainbow is of course very similar, but also will not be repeated here.

Further comments

There is an interesting optical illusion associated with a rotating lighthouse beam. The beam appears to change its length in a continuous fashion, attaining its maximum length when passing overhead, and reaching its minimum when the beam is pointing directly away from the observer. To understand why this is so, let's visit a star party.

Have you ever used a flashlight beam as a pointer to identify particular stars, star groups, or planets to interested bystanders? Perhaps there were none. No matter; you may have noticed that even if the sky is clear, the beam seems to come to an end very abruptly in a particular direction. But why do we see the beam in the first place? The explanation is the same as that for the visibility of sunbeams: dust particles and water droplets in the beam scatter the light, some of which of course enters the eyes of the observer. In Figure 23.9, the observer—you—is at the point O and the beam of light starts at L and extends (while diverging somewhat) to C and beyond. From points A, B, C, etc. in the

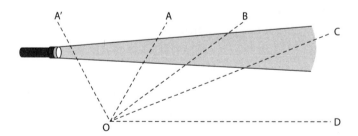

Figure 23.9. Geometry of the flashlight beam. The observer is at **O**. The point L (not shown) is where the beam begins.

forward direction, light is scattered toward the observer, but no matter how much the beam extends, the observer, displaced from it, will never see it extend beyond the direction OD (parallel to the axis of symmetry of the beam). When such a beam is rotating, the apparent length will change in accordance with the angle the beam makes with the observer-source line.

But why can we see a lighthouse beam anyway? Or a flashlight beam, for that matter? The answers may seem obvious; it's all about photons entering our eyes, isn't it? Well, yes (always) and no! If a flashlight is shone directly into our eyes, we have no choice but to see it, unless we close our eyes. But if it is pointed away from us, we can usually still see it. The closer the direction of sight is to that direction, the wider the beam is, and the more scattering particles there are, and this will to some extent counteract the fact that the beam is more remote from O in that direction. However, if you look in the direction OA' and compare the intensity of scattered light with that when you look in the direction OA (roughly 90° from OA'), you will notice that the former is considerably greater: more light is scattered in the forward direction than in the backward one. The very same phenomenon occur with lighthouse beams—a beam pointing toward the observer is brighter than when it points away from him.

As we noted in more detail in Chapter 20, the explanation is that the particles scattering the light are sufficiently large that they scatter asymmetrically: much more light is scattered in the forward direction. This is (in case you had forgotten) essentially why the color of the blue sky can change as a result of changing proportions of dust, ash, salt particles, and water droplets in the atmosphere, and it also accounts for some of the differences between a blue sky in southern Europe, or the tropics, and a blue sky (when it occurs!) in northern England. Basically, small particles (of size $\approx 10^{-8}$ m) scatter blue and violet light most, and with almost equal intensity in all directions, whereas large particles (of size $\approx 10^{-6}$ m) scatter all colors more or less equally, but mostly at small scattering angles (i.e., in the forward direction).

Some final comments can be made to wrap things up in this chapter. We have discussed rainbow and halo slices in lighthouse and searchlight beams, but can we really see the rainbow colors in these cases? Generally, it seems unlikely for several reasons, one being that the human eye is less sensitive to color at low light intensities, as evidenced by the contrast between rainbows and halos produced in moonlight as opposed to sunlight. Perhaps a more substantial reason is that modern lighthouses use sodium or mercury-vapor lights

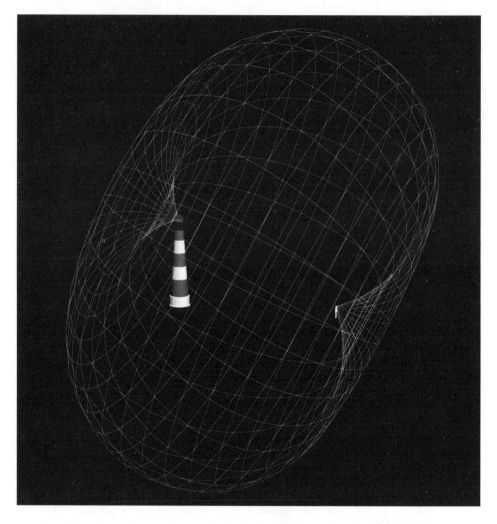

Figure 23.10. "Minnaert cigar" for a lighthouse beam. Compare the surface near the light with the shape of the rainbow arc in Figure 23.1. Courtesy of Achim Christopher.

(which are "nearly" monochromatic) or incandescent bulbs, though light from the latter can exhibit a reddish tinge sufficiently far from the source—for reasons the reader should be able to infer by now!

Lighthouse beams are generally quite divergent, unlike the more collimated versions implied by the figures above. How does this affect the conclusions we have drawn on the basis of some rather interesting trigonometry? Instead of "slices," the rainbows (or at least, the brighter portions of the beam) will take the shape of part of the (appropriate) curve shown in Figure 23.5. This is beautifully illustrated in Figure 23.2. The corresponding apple-shaped "Minnaert cigar" that we have discussed in Chapter 22 is illustrated in Figure 23.10.

Chapter 24

DISASTER IN THE CITY?

$\mathbf{H}$ ollywood is very fond of making disaster movies about volcanic eruptions, earthquakes, meteoric impacts, and alien invasions, among other
threats to us Earthlings. In July 1994 there was an alien invasion of sorts—on
the planet Jupiter. The following news flash [41] can be found on California
Institute of Technology's Jet Propulsion Laboratory (JPL) website:

> From July 16 through July 22, 1994, pieces of an object designated
> as Comet P/Shoemaker-Levy 9 collided with Jupiter. This is the first
> collision of two solar system bodies ever to be observed, and the ef-
> fects of the comet impacts on Jupiter's atmosphere have been simply
> spectacular and beyond expectations. Comet Shoemaker-Levy 9

consisted of at least 21 discernable fragments with diameters estimated at up to 2 kilometers.

In light of the impact(s) of ex-comet Shoemaker-Levy on Jupiter's outer atmosphere the question has been raised: could it happen here on earth? As opposed to a cometary encounter, we hear more these days about the possibility
of an *asteroid* colliding with the Earth. Such a collision could be globally di-
sastrous of, course, particularly if the rocky body is of the order of a kilometer
in size or more.

On May 15 1996, two University of Florida students in the astronomy
graduate program discovered an asteroid headed in the Earth's direction at
about 58,000 km/hr. As is standard procedure (and extremely sensible), they
immediately reported their observations to the Harvard-Smithsonian Center
for Astrophysics. After confirming the details—location and projected trajectory—of the object, the Center posted the relevant information about it (by
then designated 1996 AJ1) on the World Wide Web. At 4:34 p.m. (GMT) on
May 19, AJ1 reached its point of closest approach to our planet, 450,000 km,
just beyond the orbit of the moon (400,000 km). We were safe! More recently,
however, an even closer encounter occurred. On November 9, 2011, the popular site *Astronomy Picture of the Day* (http://apod.nasa.gov/apod/astropix.
html) posted a blurry picture of an asteroid with the following description:

"Asteroid 2005 YU55 passed by the Earth yesterday, posing no danger. The
space rock, estimated to be about 400 meters across, coasted by just inside the
orbit of Earth's Moon. Although the passing of smaller rocks near the Earth
is not very unusual—in fact small rocks from space strike Earth daily—a rock
this large hasn't passed this close since 1976. Were YU55 to have struck land, it
might have caused a magnitude seven earthquake and left a city-sized crater. A
perhaps larger danger would have occurred were YU55 to have struck the ocean
and raised a large tsunami.... Objects like YU55 are hard to detect because
they are so faint and move so fast. However, humanity's ability to scan the sky to
detect, catalog, and analyze such objects has increased notably in recent years."

And we were still safe!

$X = L$: THE ASTEROID PROBLEM

But there *have* been meteoric impacts during the Earth's long history. Let's revisit two such instances (without a time machine), and do a little mathematics

in the process. About 65 million years ago such an encounter occurred (Alvarez et al. 1980), and it may well have caused the demise of the dinosaurs. The site where the explosive encounter is believed to have taken place is called the Chicxulub crater; it is an ancient impact crater buried underneath the Yucatán Peninsula in Mexico. Dust from the impact was lofted into the upper atmosphere all around the globe, where it lingered for at least several months. Eventually it settled back to the surface of the earth, having done a superb job of blocking sunlight and thus devastating plant and animal life. On the dark and cold Earth that temporarily resulted, many life-forms became extinct. Available evidence suggests that about 20% of the asteroid's mass ended up as dust settling out of the upper atmosphere. This dust amounted to an average of about 0.02 gm/cm² on the surface of the Earth. If this ancient asteroid had a density of about 2 gm/cm³ (about the density of moon rock), *how large was it*?

The radius of the Earth is about 4000 miles, or 6400 km. This is 6.4×10^8 cm, so the area of the Earth's surface is about $4 \times 3 \times 40 \times 10^{16} \approx 5 \times 10^{18}$ cm². Each cm² contained, according to hypothesis, 0.02 gm of asteroid dust. The total mass that settled out was therefore about 10^{17} gm, and the mass of the asteroid was about five times this, or 5×10^{17} gm. Since the shape was almost certainly irregular, let's replace it with the cube of the same mass; the size of the cube will differ little from the average dimension of the asteroid. A cube of side L with this mass and supposed density 2 gm cm⁻³ means that $2L^3 = 5 \times 10^{17}$, or $L = (0.25 \times 10^{18})^{1/3}$ cm ≈ 6 km. This suggests that the asteroid was, to the nearest order of magnitude, about 10 km in size. This is not unreasonable for an asteroid (though the dinosaurs might well have disagreed). The impact is estimated to have released 4×10^{23} joules of energy, equivalent to 10^8 megatons of TNT (trinitrotoluene) on impact. Comparing this with equivalent TNT yield of the atomic bombs dropped on Hiroshima (12–15 kilotons) and Nagasaki (20–22 kilotons) in August 1945, one can quite see why such an impact can wreak so much devastation.

$X = M$: METEOR CRATER, WINSLOW, ARIZONA

Far more recently, between 20,000 and 50,000 years ago, the Earth had another visitor. A much smaller object with diameter between 40 m and 50 m in size hurtled toward Flagstaff, Arizona, before there was a Flagstaff, or even an Arizona. Its speed is estimated to have been about 72,000 km/hr (20,000 m/s). Made mostly of iron, its density was probably about 8000 kg/m³ (**question:**

what is this in gm/cm^3, the units used above?). A cube of side 45 m made of this material would have a mass $M \approx 8000 \times (45)^3 \approx 7 \times 10^8$ kg, or about 700,000 metric tons. Some studies suggest the object was smaller, approximately a sphere of diameter 40 m, with volume about 260,000 metric tons. Either way it is pretty large; the battleship *Iowa* by comparison displaces about 50,000 metric tons. The kinetic energy on impact for the smaller mass is $1/2 \times$ mass $\times$ (speed)$^2 \approx 5 \times 10^{16}$ joules, roughly equivalent to the energy released by a thermonuclear bomb (20 tons of TNT).

X = P: YET MORE TO BE CONCERNED ABOUT?

Suppose that the object called 1996 AJ1(comparable in size with 2005 YU55) had indeed struck the Earth...what would have been the "impact," and how does it compare with that of the Winslow meteorite? It was estimated to be about ten times the size of that one, so the volume must be roughly a thousand times greater. However, it is believed to be composed of rock and ice with an average density of about 3000 kg m^{-3}, considerably less than the chunk of iron that created the hole in Arizona. The speed at its closest approach was about 16,000m/s, or 80% of the estimate for the Winslow object. Then the kinetic energy would have been about $10^3 \times \frac{3}{8} \times (0.8)^2 \approx 240$ times larger; say between two and three hundred times larger. Since the Earth's surface is 70% water, there is a good chance it would have hit the ocean, creating huge tsunami waves. These would of course have been extremely destructive, as we know from earthquake-initiated events around the globe. Had it hit a large city such as New York or Washington, D.C., the devastation would have been widespread. To quote one source [42]:

> Utilizing scaling models presented by Shoemaker (1983), one can calculate that the crater produced by the May 19 meteor would have had a diameter $D = 8,500$ m and a depth $h = 1,200$ m. The volume of the crater would have been about 34 billion cubic meters. A direct hit on Washington, D.C. would have completely obliterated the entire central region of the city. The Potomac River would have quickly filled the crater to produce a large deep lagoon.

So what is the probability of a big chunk of rock hitting a city? It is of course extremely difficult to assess this risk, but we can obtain an upper bound (of sorts) on the probability that *if an asteroid is heading directly for the Earth* (so that impact somewhere is inevitable), it will hit a major population center. Note that this is *very* different from the probability that an asteroid will impact the Earth. According to one online source [43] (using data from January 2007) ranking the largest cities by land area, that ranked first is the New York Metropolitan area with about 9000 km² (I am rounding these area and population figures to the nearest thousand and million respectively). With a population of about 18 million according to this source, the population density was about 2000 people/km². The 100th city in this ranking is Jeddah in Saudi Arabia at about 1000 km² (this is a slight overestimate); the population in January 2007 was about 3 million, so the corresponding population density was approximately 3000 people/km². The cities in places 50 and 51 are respectively Delhi, India, and Denver, USA, with land areas about 1300 km², though the contrast between them could not be more striking. Delhi had a population of around14 million at that time; Denver's was nearly 2 million, with a much lower population density of course. The corresponding population densities are about 10,000 and 2000 people/km². We can get a "handle" on the average population density from these figures using the Goldilocks principle, noting that it will likely be between 1000 and 10,000 people/km². This gives about 3000/km². We noted in Chapter 18 that by 2007 more than half the world's population was living in cities. Taking a world population of 7 billion (declared by the United Nations as reached October 31, 2011), we estimate that the combined area of all cities is

$$\frac{7 \times 10^9 \times (1/2)}{3 \times 10^3} \approx 10^6 \text{ km}^2.$$

How does this compare with the cross-sectional area of the Earth? This is what the asteroid will "see" as it heads toward us, and with a radius of about 6400 km, this area is $\pi \times (6400)^2 \approx 10^8$ km². The ratio of the combined metropolitan area to that of the Earth's cross section is about 10^{-2}, so it's not a tiny figure, relatively speaking. So on the basis of a *very* crude calculation, it appears that *if* the asteroid is heading directly for our planet, so an impact somewhere is inevitable, there's about a 1% chance that a city will be hit. But it may not be yours! The probability of that occurring is much lower, of course. We know that about half the world's population reside in cities. City populations in general tend to be larger than 10^5 and smaller than 10^7 (excluding really large cities, of course), so we'll apply the Goldilocks principle one more time to obtain our guesstimate for an average population: one million. Dividing three billion by this number, we have three thousand such "average" cities. We have noted above that, given an impending direct collision, there's about a 1% chance that a city will be hit. Dividing 10^{-2} by three thousand indicates that the chances of your city (or mine) being hit by an asteroid heading straight for Earth may be somewhere between 10^{-5} and 10^{-6}. Does that make you feel a little less concerned? And although the surface of the Earth is about 70% covered by water, this has no bearing on our calculation; it is just the relative area occupied by city dwellers that is important here. Of course, if the impacting body is large enough, then everyone is doomed whether it hits an ocean or not.

$X = V, A$: WINSLOW REVISITED

We finish this chapter about the "somber possibility of very serious things happening" on a more light-hearted note. The Winslow crater is *now* about 170 m deep and 1250 m across. You've heard of ships in a bottle? How about building a town in a meteor crater? The base of the crater, not being entirely flat, would probably have to be leveled, but ignoring this practical concern, we assume this has been done, and calculate the base area to be just over 1 km². This is far too small to accommodate more than a small town with the population density of the New York metropolitan area (about 2000 people/ km²). Well, if not a city, how about a stadium? This would be a most appropriate place to hold *rock* concerts.

We can also use integral calculus to determine the volume and surface area of the (idealized) circular crater [42] of diameter $2a$ and maximum depth h. We suppose that it is, as it were, a portion of the inner surface of a sphere of radius r and center at the origin O (see Figure 24.1).

First, the volume:

$$V = \pi \int_{r-h}^{r} x^2\, dy = \pi \int_{r-h}^{r} (r^2 - y^2)\, dy. \qquad (24.1)$$

After a little algebra, and using the result

$$r = \frac{h}{2} + \frac{a^2}{2h}$$

we find that (**Exercise!**)

$$V = \frac{\pi h}{6} (h^2 + 3a^2). \qquad (24.2)$$

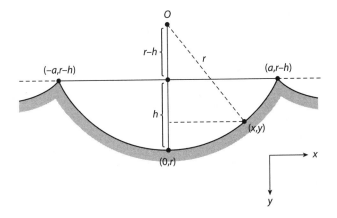

Figure 24.1. An idealized cross section of the Winslow meteor crater.

The area is found in a similar manner (**Exercise!**), thus:

$$A = 2\pi \int_{r-h}^{r} x\left(1 + \frac{y^2}{x^2}\right)^{1/2} dy = 2\pi \int_{r-h}^{r} r\,dy = \pi(h^2 + a^2).$$

So: plugging in the values $a = 625$ and $h = 170$ we find that $V \approx 1.0 \times 10^8 \, \text{m}^3$ and $A \approx 1.3 \times 10^6 \, \text{m}^2$.

Exercise: Given that the density of the ground material at the crater is about 2350 kg/m^3, *mentally* calculate the mass displaced by the meteor.

Suppose that an institute of higher education, Laid Back State University (LBSU) decides to build a football stadium in the crater. LBSU could be any such institution in the United States, of course. (The one at which I am employed is about 2000 miles from Winslow, Arizona. This is considered to be "local" on the scale of the Solar system.) Allowing 15% of the total area to consist of stairs, aisles, restrooms, and hot-dog stands and the rest for the football field, and allowing 75 cm^2 for each seat, the crater can accommodate $(0.85 \times 1.30 \times 10^6)/(0.75)^2 \approx 2 \times 10^6$ people—this is just a tad larger (a hundred times!) than the recently built football stadium at Old Dominion University, which holds just under 20,000 people at maximum capacity. We still have a way to go.

Chapter 25

GETTING AWAY FROM THE CITY

After endless days of commuting on the freeway to an antiseptic, sealed-window office, there is a great urge to backpack in the woods and build a fire.

—Charles Krauthammer

Why would one wish to get out of the city, at least for a time? To escape the possible impending doom discussed in the previous chapter? There are many other reasons why we might wish to take a break from city life, such as dissatisfaction with continued traffic congestion and the hectic pace of life, or just a desire to experience nature in all its variety. My favorite seasons are spring and fall (autumn). The season of autumn can be one of great beauty, especially where the foliage changes to a bright variety of reds, oranges, and yellows.

It's a bit of an aside, but why do leaves change their colors in the fall? There are in fact several different reasons, but the most important is the increasing length of night and cooler temperatures at night. Other factors are the amount of rainfall and the overall weather patterns in the preceding months. Just like sunsets, the weather before each fall is different. Basically the production of chlorophyll slows and stops in the autumn months, causing the green color of the leaves to disappear, and the colors remaining are mixtures of brown, red, orange, and yellow, depending on the types of tree. To have any real chance of seeing the wonderful fall foliage, you have to go to the right places at the right time. Going to the beach in summer or even the fall won't do! And going to the Blue Ridge Mountains or New England in the depths of winter will not enable you to see the fall foliage either, pretty though the snow-covered trees may be!

But there are some other aspects of this season that are present at any time of the year, and do not depend on the leaves changing. Those aspects involve trees, rain, and, in this case, my left foot!

Consider this: you are in the hills of New Hampshire, or West Virginia, or perhaps you are somewhere on the Appalachian Trail enjoying the glorious fall colors, when suddenly a rain squall appears out of nowhere (or so it seems). You run to take cover in a deserted shelter a hundred yards away near the trail. Playfully (there being nothing else to do) you stick your foot outside the shelter, and of all things, photograph the rain falling around it!

After ten minutes of intense rainfall, it stops as suddenly as it started. You wonder how fast that rain must have been falling to create the scene you now survey: large puddles all along the trail, drops dripping from every available leaf above you, and the temporary dark-brown stains on the trunks of rain-soaked silver birches.

As you set out on your way again, you start to notice the wave patterns formed when the drops falling from the branches above you hit the surface of puddles.

Question: Which of the patterns in Figures 25.1 and 25.2 represents this situation?

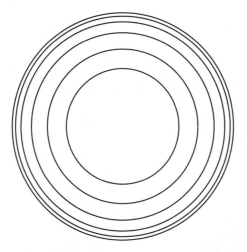

Figure 25.1. Circular wave pattern I—caused by raindrops?

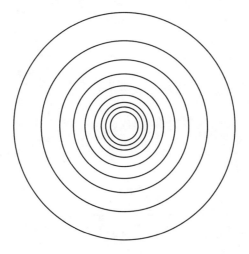

Figure 25.2. Circular wave pattern II—caused by raindrops?

Answer: The one in Figure 25.1. Raindrops falling on the surface of a puddle generate wave patterns that are dominated by the effects of surface tension. The speed of these waves is *inversely* proportional to the square root of the wavelength; thus *shorter* waves travel faster and move out first. Note the expanding region of calm associated with and inside these waves. The other

Figure 25.3. This rather pedestrian photograph (!) was taken from a sheltered area outside the Mt. Washington Resort, Bretton Woods, New Hampshire, on September 30, 2010, courtesy of Tropical Storm Nicole. Because of the heavy rain, Mt. Washington was nowhere to be seen! Nevertheless, it seemed like fun to "guesstimate" the speed of the raindrops, given that the exposure time for the shot was 1/200 second and using an estimate for the width of my sneaker (you don't need to know my shoe size to do this.). My foot and the raindrops shown were about the same distance from the camera. You can assume that the foreshortening of the rain streaks (due to the downward angle of the camera) is not significant.

pattern is dominated by gravity, which produces longer waves with speeds *directly* proportional to the square root of the wavelength, so the *longer* waves travel faster and move out first.

Once home, you upload your pictures onto the computer. On noticing the picture you took back at the shelter, you realize that knowing the exposure time of the shot, you can estimate the speed of the rain. The picture is given in Figure 25.3; can you find the speed of the rain?

Solution: From the photograph, $\frac{\text{Streak length}}{\text{Foot width}} \approx \frac{1}{3}$. The width of my sneaker is approximately 5 inches, so the raindrops traveled approximately 5/3 inches in 1/200th of a second, or 1000/3 inches/s, or 1000/36 ft/s. Therefore

$$\frac{1000 \text{ ft}}{36 \text{ s}} = \frac{1000 \text{ ft}}{36 \text{ s}} \times \frac{1 \text{ mile}}{5280 \text{ ft}} \times \frac{3600 \text{ s}}{1 \text{ hr}} \approx \frac{10^5}{5280} \text{ mph} \approx 19 \text{ mph}.$$

Since 1 m/s $\approx$ 2.2 mph, this is about 8.6 m/s.

For those who don't have access to my foot, anything in the range 4 inches to 6 inches wide is a reasonable estimate. This would result in a range of speeds 15–23 mph (7–10 m/s).

THEOREMS FOR PRINCESS DIDO

Being a princess isn't all it's cracked up to be.

—Princess Diana

We shall state two theorems that undergird Princess Dido's "solution" discussed in Chapter 1: (i) the isoperimetric property of the circle (without proof), and (ii) the "same area" theorem (with proof). The word *isoperimetric* means "constant perimeter," and here concerns the largest area that can be enclosed by a closed curve of constant length. We shall be a little more formal in our language for the statement of these results.

Theorem 1 (the isoperimetric theorem): Among all planar figures of equal perimeter, the circle (and only the circle) has maximum area.

There is also an equivalent "dual" statement:

Theorem 2 (the same area theorem): Among all planar figures of equal area, the circle (and only the circle) has minimum perimeter.

This theorem is a consequence of Theorem 1, as we can show. Consider first a closed curve C of perimeter L enclosing a domain of area A. Let r be the radius of a circle of perimeter L: then $L = 2\pi r$. This encloses a disk of area πr^2. By the isoperimetric property of the circle, the area enclosed by C cannot exceed πr^2, and it equals πr^2 if and only if C is a circle. On the other hand, $\pi r^2 = L^2/4\pi$, so we obtain the *isoperimetric inequality* $A \leq L^2/4\pi$, with the equality holding *only* for a circle. Now we are ready for the proof of Theorem 2 assuming the isoperimetric theorem. Consider next an arbitrary planar figure with perimeter L and area A. Let D be a circular disk with the same area A and perimeter l: then $A = l^2/4\pi$. If $l > L$, the isoperimetric inequality would be violated, so $L \geq l$.

To complete the argument for Princess Dido and the semicircular area discussed in Chapter 1, we can examine a generalization of this problem. Suppose that a curve of length L is now attached to a line segment of fixed length l, forming a closed curve. What should be the shape of L in order for the area enclosed to be a maximum? Consider two situations, the first in which L is a circular arc (enclosing an area A_c), and the second in which it is not (and enclosing an area A). Imagine also that in both cases a circular arc below the line segment l is added to give an additional segment of area a. By the isoperimetric theorem, $A + a \leq A_c + a$, so that $A \leq A_c$; clearly Princess Dido knew what she was doing!

PRINCESS DIDO AND THE SINC FUNCTION

We'll make an alliterative foray further into this forest by generalizing Dido's "classical" Carthage problem somewhat. Recall that she formed a semicircular arc, with the (presumed) straight Mediterranean coastline as a diameter, thus enclosing a semicircular area for the city.

Clearly, the strip length L was shorter than the Mediterranean coastline, but what about the case where the available straight boundary is of *variable* length $l < L$; how does the area of the closed region vary with l? In fact, how do we know now that the semicircular area is still a maximum? We'll do a little more mathematics and put in some numerical values just for fun, and work with l expressed in units of L, so that $0 < l < 1$. Note that if $l = 1$ the enclosed area A is zero; the strip $L = 1$ is collinear with l, and if $l = 0$ the area of the circle is just $1/4\pi$ square units, half that of the area enclosed by the boundary and the completing semicircular arc. To investigate this, examine the circular segment BCD in Figure A2.1. The shaded area $A(l)$ is enclosed by the circular arc of fixed unit length and the line segment of length l.

The arc subtends an angle 2θ at the center O, $0 < \theta \leq \pi$. The arc length $2r\theta = 1$. In terms of θ the area of the shaded segment is

$$A(\theta) = r^2(\theta - \sin\theta\cos\theta) = \left(\frac{1}{2\theta}\right)^2\left(\theta - \frac{\sin 2\theta}{2}\right). \qquad (A2.1)$$

Note that, for $\pi/2 < \theta \leq \pi$ this formula is still correct, but now $\cos\theta < 0$. A graph of $A(\theta)$ is shown below (Figure A2.2). It does indeed appear that the maximum enclosed area occurs when the region is a semicircle ($\theta = \pi/2$). This is easily verified by noting that

$$\frac{dA}{d\theta} = \frac{\cos\theta}{2\theta^2}\left(\frac{\sin\theta}{\theta} - \cos\theta\right)$$

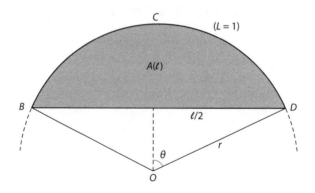

Figure A2.1. Sector notation.

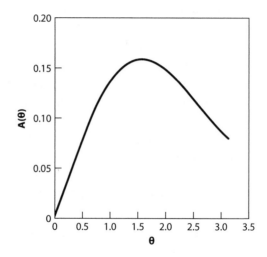

Figure A2.2. Area of sector as a function of θ (radians).

vanishes when $\theta = \pi/2$ and $\theta \approx 4.49$ (the latter is well outside the interval of interest). And

$$A(\pi/2) = 1/2\pi \approx 0.16.$$

But what is $A(l)$? Although (as indicated below) we cannot write down an exact expression for $A(l)$, we can do so for $A'(l)$, which is even better, for we can find values of θ corresponding to an extremum of A. Since

$$l = 2r \sin \theta = \frac{\sin \theta}{\theta} \tag{A2.2}$$

and

$$\frac{dA}{dl} = \left(\frac{dA}{d\theta}\right)\left(\frac{d\theta}{dl}\right)$$

we readily obtain the expression

$$\frac{dA}{dl} = \frac{\cos \theta}{2\theta^3}\left[\left(\frac{\sin \theta}{\theta}\right)^2 - \cos^2 \theta\right],$$

which is zero when $\theta = \pi/2$ or $\tan \theta = \pm \theta$. (Note that dA/dl is expressed as a function of θ.) A graph of this derivative over the interval $(0, \pi)$ shows that it is positive in $(0, \pi/2)$, zero at $\theta = \pi/2$, negative in $(\pi/2, 2.029)$ and positive once more in $(2.029, \pi)$. Thus $A(l(\theta))$ has a maximum at $\theta = \pi/2$ and a minimum at $\theta \approx 2.029$. When $\theta = \pi/2$, $A = 1/2\pi \approx 0.159$ and $l = 2/\pi \approx 0.64$.

In using the result $l = (\sin \theta)/\theta$ we have made a natural connection between the two independent variables used above to express the area of the enclosed region. More precisely, the function

$$\begin{aligned} \mathrm{sinc}\,(\theta) &= \frac{\sin \theta}{\theta}, \theta \neq 0; \\ \mathrm{sinc}\,(\theta) &= 1, \theta = 0 \end{aligned} \tag{A2.3}$$

is known as the *sine cardinal function*, shown in Figure A2.3 for the interval $(-4\pi, 4\pi)$.

The sinc function arises in many applications, from diffraction theory in optics to spectroscopy, information theory, and digital signal processing.

Unfortunately, there is no nice "inverse sinc function" to which we can appeal to express θ directly in terms of l. Nevertheless, in an entrepreneurial spirit, we shall attempt to find a reasonable and invertible approximation to this function, sufficient for our purposes. Since we are only interested in the interval $[0, \pi]$ we shall use instead the function

$$f(\theta) = 1 - \left(\frac{\theta}{\pi}\right)^2. \tag{A2.4}$$

The graphs of these two functions are shown in Figure A2.4; $f(\theta)$ (the solid line) appears to be good enough for our purposes!

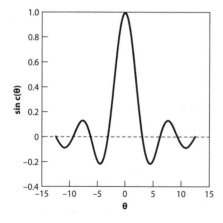

Figure A2.3. The sinc function.

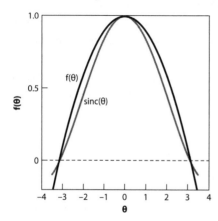

Figure A2.4. The approximation $f(\theta)$.

In terms of this approximate function, it follows that since $l \approx f(\theta)$, then

$$\theta \approx \pi\sqrt{1-l}, \tag{A2.5}$$

and hence from equation (A2.1),

$$A(l) \approx \frac{\beta(l)-\sin\beta(l)}{2\beta^2(l)} = \frac{1-\operatorname{sinc}\beta(l)}{2\beta(l)}, \tag{A2.6}$$

where $\beta(l) = 2\pi\sqrt{1-l}$. The approximation (A2.6) for $A(l)$ is illustrated in Figure A2.5.

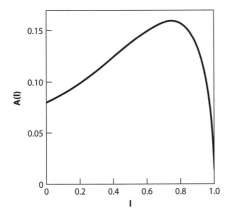

Figure A2.5. The approximation for **A(l)**.

How does this compare with the exact results? The maximum value of $A(l)$ according to equation (A2.6), occurs when

$$\frac{dA}{dl} = \left(\frac{dA}{d\beta}\right)\left(\frac{d\beta}{dl}\right) \approx -\frac{\pi}{\sqrt{1-l}}\frac{\cos(\beta/2)}{\beta^2}\left[\frac{\sin(\beta/2)}{\beta/2} - \cos(\beta/2)\right] = 0.$$

The only value of interest to us is when $\beta = \pi$, which corresponds to $A \approx 0.159$, $\theta = \pi/2$ and $l = 0.75$. Therefore we see that the approximation is "spot on" for the maximum area and where it occurs, but it overestimates the corresponding value of l by about 17%.

Comment: The definition (A2.3) is sometimes called the unnormalized sinc function. The *normalized* sinc function is defined as

$$\mathrm{sinc}\,(\theta) = \frac{\sin \pi\theta}{\pi\theta}, \theta \neq 0;$$
$$\mathrm{sinc}\,(\theta) = 1, \theta = 0$$

because it may be shown that

$$\int_{-\infty}^{\infty} \frac{\sin \pi\theta}{\pi\theta}\, d\theta = 1.$$

TAXICAB GEOMETRY

> If, in New York, you arrive late for an appointment, say, "I took
> a taxi."
>
> —*Andre Maurois*

I magine a city laid out on a standard rectangular (Cartesian) grid, with blocks defined by equally spaced N-S and E-W roads (of course any directions will do as long as they are perpendicular). Thus to move from location P to location Q we must travel along these roads; we cannot generally move in a direct line from P to Q unless we happen to be a pigeon! Let's compare some of the properties of this somewhat "constraining" geometry with the standard Euclidean one available for the pigeons. Find some graph paper and let's get started!

Suppose for example that P is the point $(1, 3)$, my office, and Q is the point $(-2, -1)$, where Starbucks is located. We are of course assuming that the size of each business is small compared with a block distance, so that we can represent them by points. The Euclidean distance between them, $d_E(PQ)$ is readily computed from the distance formula:

$$d_E(PQ) = \sqrt{[1-(-2)]^2 + [3-(-1)]^2} = \sqrt{9+16} = 5 \text{ blocks.}$$

(From now on we will drop the unit of distance "blocks": it will be understood.) But of course, to arrive at Starbucks from my office I must travel either from P to $A(-2, 3)$ and then to Q, or from P to $B(1, -1)$ and then to Q. In each case the "taxicab" distance is given by

$$d_T(PQ) = |1-(-2)| + |3-(-1)| = 7.$$

Clearly, $d_T(PQ) > d_E(PQ)$ in this case. But inequalities between distances in one geometry are not necessarily preserved in the other. Consider, for example, the Euclidean distances between (i) the points $R(1,1)$ and $S(3,3)$, and (ii) $R(1,1)$ and $T(-2,1)$, which are, respectively,

$$d_E(RS) = \sqrt{8} \text{ and } d_E(RT) = 3, \text{ so that } d_E(RT) > d_E(RS).$$

However,

$$d_T(RS) = 4 \text{ and } d_T(RT) = 3, \text{ so that } d_T(RT) < d_T(RS).$$

This non-Euclidean geometry, informally known as *taxicab geometry* (see [44]) is a form of geometry in which the usual measure of distance in Euclidean geometry is replaced by a new one in which the distance between two points is the sum of the (absolute) differences of their coordinates. This new measure of distance (or "metric") is also sometimes known as the *Manhattan distance*; in the plane, the Manhattan distance between the point P_1 with coordinates (x_1, y_1) and the point P_2 at (x_2, y_2) is $|x_1 - x_2| + |y_1 - y_2|$. Taxicab geometry satisfies all Euclid's axioms except for the side-angle-side axiom, because one can generate two triangles with two sides of the same length in this metric, and with the angle between them the same and yet they are *not* congruent. For more information on the sometimes surprising properties of this "urban geometry", consult the delightful little book on this topic by Krause (1986).

Question: Is Pi equal to four?

Well, as they say, it all depends. . . . If we sketch the set of all points that are a unit Euclidean distance from a fixed point (the coordinate origin, without loss of generality), then we have the unit circle. If we do the same thing but in the taxicab "metric"as it is called, what shape do we get? A little thought, coupled with a sketch or two, reveals that we get a "diamond" shape, that is, a square with two vertices lying on the "y" axis and the other two on the "x" axis. This is a *taxicab circle* of unit radius! Now we are familiar with the definition of π as the circumference of a circle divided by its diameter. With the taxicab metric, the "length" of one side of our taxicab circle is two units, hence the circumference is eight units, and the diameter is two units, so, yes, the taxicab value of "taxicab π" is indeed four!

Of course, there are potentially an infinite number of distance metrics and corresponding "circles." The metric (or formula) for the distance of a point (x,y)

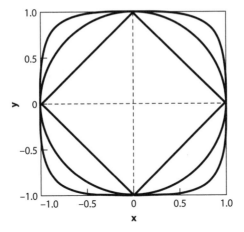

Figure A3.1. Taxicab "circles" d_1, d_2, d_4, and d_∞.

from the origin in Euclidean geometry is of course well known: $d_2 = \sqrt{x^2 + y^2}$. This is readily generalized for other positive integers n such that $d_n = \sqrt[n]{x^n + y^n}$. The case for taxicab geometry corresponds to $n = 1$, for which $d_2 = |x| + |y|$. As n increases in value, the "circles" become more and more bulbous, as seen in Figure A3.1 for several values of n, and in the limit as $n \to \infty$, the metric is defined to be $d_\infty = \max\{|x|, |y|\}$. As can be seen from the figure, the unit circle for the d_∞ metric is a square with sides parallel to the coordinate axes.

Exercise 1: Mark the points $A = (-2, -1)$ and $B = (2, 2)$ on a piece of graph paper. Then sketch the locus of all points P such that $d_T(P,A) + d_T(P,B) = 9$. By analogy with the corresponding figure in Euclidean geometry provide a name for this figure.

Exercise 2: Mark the points $A = (-3, -1)$ and $B = (2, 2)$ on a piece of graph paper. Then sketch the locus of all points P such that $|d_T(P,A) - d_T(P,B)| = 3$. Again, by analogy with the corresponding figure in Euclidean geometry provide a name for this figure.

THE POISSON DISTRIBUTION

Applications of the Poisson distribution are many and varied. A large class of natural and social phenomena have been successfully modeled using it [45]. Before deriving the formula for this distribution let's consider some examples. A classic one concerns a Russian statistician, Ladislaus Bortkiewicz, who used the Poisson distribution to estimate the number of soldiers killed by mule-kicks to the head in the Prussian army. He assembled data on the 14 cavalry corps over a period of 20 years, and in so doing he was able to verify that the distribution of mule-kick deaths fit a Poisson distribution! Another example is the number of armadillos killed by traffic on a length of an Arizona highway; this is also "Poisson distributed." Other examples are the number of emergency room cases cases arriving at a hospital during a one-hour period, and the number of bomb hits in one-acre areas of metropolitan London during World War II [45]. More pedestrian examples include the number of times I check my email in the morning, the number of cars that pass a particular "eatery" in a two-hour period (perhaps the "*Road-Kill Café*" in a certain Arizona town?).

This is not of course to imply that everything can or should be modeled as a Poisson process, however; there are certain restrictions that must be satisfied before the phenomenon of interest can be thus described. The ER cases arriving on a Sunday morning would probably not have the same distribution as those on a Saturday evening, for example. Also, if there were an explosion or tornado, the arrivals would be associated with a common cause, and this could render the Poisson distribution inappropriate. If the area of London were extended too far into the surrounding countryside, the intensity of the bomb damage would be much less severe. The armadillos are presumed to cross the highway at random locations, not to be traveling as a herd, or in convoy, or crossing the road at a particular spot (as Canada geese seem to do, especially when I am in a hurry).

Perhaps the simplest context in which to place the mathematical discussion is found by asking the following question. What is the probability of getting exactly n tails in N tosses of a (possibly unfair) coin, where $n \ll N$? This means that there will be $N - n$ heads, of course, and if the probability of getting a tail is p, then that of getting a head is $1 - p$. Of course, if the coin is fair, then $p = 1/2$. The probability of n tails is therefore the number of ways one can have n tails in N tosses multiplied by the probabilities associated with each toss. The result is

$$\Pr(N, n) = \frac{N!}{n!(N-n)!} p^n (1-p)^{N-n}. \tag{A4.1}$$

This is known as the binomial probability distribution. To obtain the Poisson distribution as a limiting case of this distribution we suppose getting a tail is a very rare event, that is, $p \ll 1$; this corresponds to a very unfair coin! We can (eventually) simplify expression (A4.1). First, we note that for $p \ll 1$, $\ln(1-p) \approx p$, so that

$$(1-p)^{N-n} = e^{\ln[(1-p)^{N-n}]} = e^{(N-n)\ln(1-p)} \approx e^{-(N-n)p} \approx e^{-Np}. \tag{A4.2}$$

Noting that

$$\ln\left[\frac{N!}{(N-n)!}\right] = \ln(N!) - \ln((N-n)!), \tag{A4.3}$$

we can approximate the logarithm of the factorial $M!$ for a large positive integer M by a "quick and dirty" method as follows:

$$\ln M! = \sum_{k=1}^{M} \ln k \approx \int_1^M \ln x \, dx = [x \ln x - x]_1^M = M \ln M - M + 1$$
$$\approx M \ln M - M. \tag{A4.4}$$

From this it follows directly that $M! \approx M^M e^{-M}$. This result is sometimes known as the weak form of *Stirling's approximation*. Returning to equation (A4.3) and applying it to the two terms on the right of equation (A4.4), it follows that

$$\ln\left[\frac{N!}{(N-n)!}\right] = N \ln N - N - (N-n) \ln(N-n) + (N-n).$$

Since $n \ll N$,

$$\ln(N - n) = \ln N + \ln(1 - n/N) \approx \ln N - n/N.$$

Therefore

$$\ln\left[\frac{N!}{(N-n)!}\right] = n(\ln N - n/N) \approx n \ln N.$$

We have neglected the term in n^2 since N is large. Hence

$$\frac{N!}{(N-n)!} \approx N^n. \tag{A4.5}$$

Therefore, combining the approximations (A4.2) and (A4.5), we have

$$\Pr(N,n) \approx \frac{(Np)^n e^{-Np}}{n!}. \tag{A4.6}$$

The expected number of tails in N tosses is $\lambda = Np$. Thus equation (A4.6) can be written in terms of the expectation λ as

$$\Pr(N,n) = \frac{\lambda^n e^{-\lambda}}{n!}. \tag{A4.7}$$

This is the Poisson distribution for the probability of exactly n tails occurring in N tosses of the coin, when $p \ll 1$. And instead of heads and tails, we can think of customers arriving at the post office, or gaps in traffic, as applied in Chapters 3 and 9, respectively.

We can think of this in another way. Note that

$$1 = e^{\lambda} e^{-\lambda} = \left(\sum_{n=0}^{\infty} \frac{\lambda^n}{n!}\right) e^{-\lambda} = \left(1 + \frac{\lambda}{1!} + \frac{\lambda^2}{2!} + \frac{\lambda^3}{3!} + \ldots\right) e^{-\lambda}$$
$$= \Pr(N,0) + \Pr(N,1) + \Pr(N,2) + \Pr(N,3) + \ldots$$

This result can be interpreted as the sums of the probabilities of all possible outcomes occurring; in other words, we have a probability distribution.

THE METHOD OF LAGRANGE MULTIPLIERS

We state without proof a rather formal expression of a general problem to which this method applies:

Any local maxima or minima of a function $z = f(x,y)$ that is subject to the constraint $g(x,y) = 0$ will be among those points (x_0, y_0) for which the point (x_0, y_0, λ_0) is a solution to the system of equations

$$\frac{\partial F}{\partial x} = 0, \frac{\partial F}{\partial y} = 0, \text{ and } \frac{\partial F}{\partial \lambda} = 0,$$

where

$$F(x,y,\lambda) = f(x,y) + \lambda g(x,y),$$

provided these partial derivatives exist.

Appendix 6

A SPIRAL BRAKING PATH

What is the path of a particle that is acted upon by a constant force? Since force is a vector quantity, this means that both the magnitude a and the direction of the force are constant. Suppose that the acceleration vector $\bar{a}$ makes an angle γ with the trajectory, as indicated in Figure A6.1. Many calculus textbooks demonstrate that the components of acceleration (v being speed) along the tangent and normal to the path are, respectively,

$$(i)\ \frac{d^2s}{dt^2} = \frac{dv}{dt} = v\frac{dv}{ds} \text{ and } (ii)\ \frac{v^2}{\rho},$$

where ρ is the (local) radius of curvature of the path. Resolving the acceleration in these directions yields the following equations:

$$v\frac{dv}{ds} = a\cos\gamma \text{ and } \frac{v^2}{\rho} = a\sin\gamma. \tag{A6.1}$$

In these expressions $a = |\bar{a}|$. Integrating the first equation, and noting that $\rho = |ds/d\psi|$. gives the result

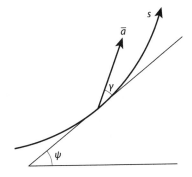

Figure A6.1. Path of the particle; ψ is the angle the tangent line makes with the polar axis, the arc length is s and the acceleration is a constant vector $\bar{a}$.

$$\frac{1}{2}\frac{ds}{d\psi} = s\cot\gamma + B, \tag{A6.2}$$

where B is a constant and we have assumed (without loss of generality) that $ds/d\psi > 0$. A further integration yields

$$\ln|s\cot\gamma + B| = 2\psi\cot\gamma + D, \tag{A6.3}$$

D being another constant. Again, without loss of generality, assuming the argument of the logarithm is positive we find that the equation of the path is

$$s = -B\tan\gamma + Fe^{2\psi\cot\gamma},$$
where $F = e^D\tan\gamma$ is yet another constant. $\tag{A6.4}$

Therefore the path is that of an equiangular spiral because (neglecting additive and multiplicative constants) it is of the form $s = e^{\text{constant}\times\psi}$.

Appendix 7

THE AVERAGE DISTANCE BETWEEN

TWO RANDOM POINTS IN A CIRCLE

Suppose the circle has radius a. We can ask: what is the joint probability that one of the points (P) is in the distance interval $(x, x + dx)$ from the center O, the other point (Q) being nearer the center, and such that PQ makes an angle with PO in the interval $(\phi, \phi + d\phi)$? From Figure A7.1, note that the probability of P being in the shaded annular region is

$$P_a = \frac{2\pi x dx}{\pi a^2}. \tag{A7.1}$$

This must be coupled with the probability of Q being in the triangular shaded sector (see Figure A7.2),

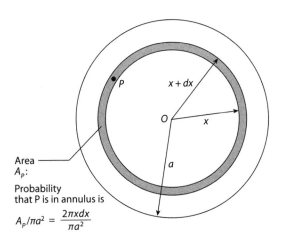

Figure A7.1. Geometry of the probability of a point P being in the annular region.

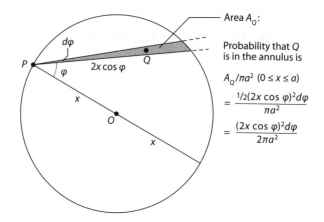

Figure A7.2. Probability of a point Q being in the (approximately) triangular sector.

$$P_s = \frac{(1/2)\,(2x\cos\phi)^2\,d\phi}{\pi a^2} = \frac{(2x\cos\phi)^2}{2\pi a^2}\,d\phi. \qquad (A7.2)$$

We also need to know the average distance between P and Q in this sector; a simple "center of mass" argument will suffice here—it lies 2/3 of the way from the "base" of the triangular sector, so the average distance is $4x(\cos\phi)/3$. Combining all these results gives expected distance between the two random points P and Q:

$$\bar{s} = \frac{32}{3\pi a^4}\int_0^a x^4\,dx \int_{-\pi/2}^{\pi/2} \cos^3\phi\,d\phi. \qquad (A7.3)$$

There is an additional factor of two in this integral to include the case when P is nearer to the center than Q. This integral can be evaluated by careful calculus students to be

$$\bar{s} = \frac{128a}{45\pi} \approx 0.905a. \qquad (A7.4)$$

Appendix 8

INFORMAL "DERIVATION" OF THE LOGISTIC
DIFFERENTIAL EQUATION

It has been said that there are two categories of people in the world: those that separate people into two categories and those that don't. Nevertheless, suppose there are indeed two categories of people (described below) comprising a community: in which the population is constant. This may seem very unrealistic, except over short periods of time, for what about births, deaths, "emigration," and "immigration" within such a community? And while this is a valid criticism, there are several "communities" that possess a constant population: cruise liners, and political bodies (House, Senate, etc.) being two such examples. We shall examine the mathematical genesis of the logistic equation in the former context, but not before identifying a limitation inherent in this approach. A cruise liner may have a thousand passengers (and several hundred crew members) on board, whereas the United States Senate has one hundred members. In each case, the total population is constant (ignoring people falling overboard in the former case, of course!), but within each community, there may be two classes of subpopulations. Obviously the number of males and females remains fixed, so what could vary?

We'll make some simplifying assumptions here. Suppose that a passenger with a contagious and easily transferable disease boards the cruise liner (without exhibiting any symptoms at that time). Over time, assuming all the other passengers are susceptible to this disease, as the infected individual comes into contact with them, the number of infected passengers increases—and this is certainly something that has happened on several occasions in recent years. So in this case, at any given time in this simplified model, there are two categories of passengers: those who are infected and those who are not. And in this model these populations will vary monotonically over time subject only to the condition that their sum is a constant, K. We could change the scenario from transference of a disease to that of a rumor: gossip! In that case there would be additional assumptions to be made: everyone who knew the rumor would be willing to share it, and everyone who did not know it would be willing to listen (and hence pass it on!). A related context is that of advertising by word of mouth: "Did you hear about the special offer being made at Sunbucks? They're giving away a Caribbean cruise to everyone who buys a Grande peppered Latvian pineapple-cauliflower espresso mocha latté."

In the case of the U.S. Senate, suppose that Senator A introduces a bill (maybe to restrict the availability of the above coffee at Sunbucks because of its harmful effects on the local populace?). Perhaps there is little support for the bill initially (many of the senators like that coffee), and as acrimonious but eloquent debate continues, more and more senators begin to see the error of their ways, and eventually vote accordingly . . .. Again, there will be an evolution of the potential "Aye" and "Nay" votes over time, culminating in, well, you'll just have to tune in to C-Span to see the outcome.

So what is the limitation of this approach, regardless of context? Nothing yet, but if we use calculus to try to describe the rate of change of the two populations, we are making the implicit assumption of differentiability, and hence continuity of the populations. But the populations are discrete! The are always an integral number of infected passengers, or of senators disposed to vote for the bill (and despite one's personal misgivings about Senator B, though he may only do the work of half a senator, he is one person). Calculus is strictly valid when there is a continuum of values for the variables concerned, and in that sense calculus-based models of discrete systems can never be totally realistic, even when there are billions of individuals (such as the number of cells in a tumor). It is usually the case in practice that the more individuals there are in a population, the more appropriate the mathematical description will be from a continuum perspective, because small populations can be subject to fluctuations that are comparable in size with the population! Under those circumstances a discrete approach is desirable. Nevertheless, when the number of possible "states" is limited (as in "Aye" or "Nay," infected or not), frequently the calculus-based approach is sufficiently accurate to "interpolate" the behavior of the more accurate discrete formulation. And that is what we shall do here in the context of the cruise liner epidemic, neglecting all complications like incubation times, likelihood of recovery, and immunity from the disease.

Such considerations are very important in realistic epidemiological models, but will not be addressed here. We are therefore assuming that all passengers who are uninfected at the beginning of the cruise are susceptible to the disease, and that once they have contracted it, they remain infected (and infectious) for the duration of the cruise (a very unfortunate scenario). The problem then becomes one of determining how the number of passengers contracting the disease changes over time. Let $N(t)$ be the number of infected passengers and $M(t)$ be the number not infected. Obviously $M(t) + N(t) = K$ and, as stated above, $N(0) = 1$ (this initial condition can be changed without loss of generality). Remembering that the argument we are making here is not a rigorous one, let us take two such passengers at random, and ask what the probability of the spread of the disease will be from such an encounter. Certainly, the only way the disease will spread is if each of these two individuals is in a different category. To reframe this in the "rumor" category, if our two individuals know of the rumor, then to begin with, they may just talk about the weather (especially if they are from the UK!); if both know the rumor, they may just talk about ... the weather. It is only when one of the two knows the rumor, and the other does not, that the rumor will spread. And so it is with the disease under the simplifying assumptions we have made. The probability of picking an individual from the infected population and then one from the uninfected is

$$P(N,M) = \frac{N}{K} \cdot \frac{M}{K-1} = P(M,N), \qquad (\text{A8.1})$$

so the probability of the infection spreading from this encounter is proportional to NM, and this is just $N(K - N)$. Furthermore, it is reasonable to suppose that the rate of change of N is proportional to this probability, that is,

$$\frac{dN}{dt} = kN(K - N), \qquad (\text{A8.2})$$

which is just the form of equation (15.8) with some obvious notational changes.

Appendix 9

A MINISCULE INTRODUCTION TO FRACTALS

The person most often associated with the discovery of fractals (and rightly so) is the mathematician Benoit B. Mandelbrot. Indeed, one doesn't need to be the proverbial rocket scientist to see why the *Mandelbrot set* is so named (for details of this amazing set, just Google the name!). The mathematics underlying the structure of fractals (geometric measure theory), however, had been developed long before the "computer revolution" made possible the visualization of such complicated mathematical objects. In the 1960s Mandelbrot pointed out some interesting but very surprising results in a paper entitled "How long is the coastline of Britain?" published posthumously by the English meteorologist Lewis Fry Richardson. That author had noticed that the measured length of the west coast of Britain depended heavily on the scale of the map used to make those measurements: a map with scale 1:10,000,000 (1 cm being equivalent to 100 km) has less detail than a map with scale 1:100,000 (1 cm equivalent to 1 km). The more detailed map, with more "nooks and crannies," gives a larger value for the coastline. Alternatively, one can imagine measuring a given map with smaller and smaller measuring units, or even walking around the coastline with smaller and smaller graduations on our meter rule. Of course this presumes that at such small scales we can meaningfully define the coastline, but naturally this process cannot be continued indefinitely due to the atomic structure of matter. This is completely unlike the "continuum" mathematical models to which we have referred in this book, and in which there is no smallest scale.

Richardson also investigated the behavior with scale for other geographical regions: the Australian coast, the South African coast, the German Land Frontier (1900), and the Portugese Land Frontier. For the west coast of Britain in particular, he found the following relationship between the total length s in km and the numerical value a of the measuring unit (in km, so a is dimensionless):

$$s = s_1 a^{-0.22}, \tag{A9.1}$$

where s_1 is the length when $a = 1$. Clearly, as a is reduced, s increases! If the measuring unit were one meter instead of one km, the value of s would increase by a factor of about 4.6 according to this model. Clearly, the concept of length in this context is rather amorphous; is there a better way of describing the coastline? Can we measure the "crinkliness" or "roughness" or "degree of meander" or some other such quantity? Mandelbrot showed that the answer to this question is yes, and the answer is intimately connected with a generalization of our familiar concept of "dimension." That is the so-called *topological* dimension, expressed in the natural numbers $0,1,2,3,\ldots$ (there is no reason to stop at three, by the way). It turns out that the concept of fractal dimension used by Mandelbrot (the *Hausdorff-Besicovich* dimension), being a ratio of logarithms, is not generally an integer. Mandelbrot defines a fractal as "a set for which the Hausdorff-Besicovich dimension strictly exceeds the topological dimension."

Consider the measurement of a continuous curve by a "measuring rod" of length a. Suppose that it fits N times along the length of the curve, so that the measured length $L = Na$. Obviously then, $N = L/a$ is a function of a, that is, $N = N(a)$. Thus if $a = 1$, $N(1) = L$. Similarly, if $a = 1/2$, $N(1/2) = 2L$, $N(1/3) = 3L$, and so on. For fractal curves, $N = La^{-D}$, where $D > 1$ in general, and it is called the *fractal dimension*. This means that making the scale three times as large (or a one third of the size it was previously) may lead to the measuring rod fitting around the curve more than three times the previous amount. This is because if $N(1) = L$ as before,

$$N\left(\frac{1}{3}\right) = \left(\frac{1}{3}\right)^{-D} L = (3^D)L > 3L \text{ if } D > 1. \tag{A9.2}$$

In what follows below we shall use unit length, $L = 1$ when $a = 1$ without loss of generality. Given that $N = a^{-D}$ it follows that

$$D = \frac{\log N}{\log(1/a)}. \tag{A9.3a}$$

More precisely, Mandelbrot used the definition of the fractal dimension as

$$D = \lim_{a \to 0} \frac{\log N}{\log(1/a)}. \tag{A9.3b}$$

If this has the same value at each step, then the former definition is perfectly general. Let us apply the definition to what has come to be called the *Koch snow-flake curve*. The basic iteration step is to take each line segment or side of an equilateral triangle, remove the middle third, and replace it by two sides of an equilateral triangle (each side of which is equal in length to the middle third), so now it is a "Star of David." Each time this procedure is carried out the previous line segment is increased in length by a factor 4/3. Thus $a = 1/3$ and $N = 4$ (there now being four smaller line segments in place of the original one), so

$$D = \frac{\log 4}{\log 3} \approx 1.26186.$$

The limiting snowflake curve ($a \to 0$) thus "intrudes" a little into the second dimension; this intrusion is indicated by the degree of "meander" as expressed by the fractal dimension D. This curve is everywhere continuous but nowhere differentiable! The existence of such curves, continuous but without tangents, was first demonstrated well over a century ago by the German mathematician Karl Weierstrass (1815–1897), and this horrified many of his peers. Physicists, however, were more welcoming; Ludwig Boltzmann (1844–1906) wrote to Felix Klein (1849–1925) in January 1898 with the comment that such functions might well have been invented by physicists because there are problems in statistical mechanics "that absolutely necessitate the use of non-differentiable functions." He had in mind, no doubt, Brownian motion (this is the constant and highly erratic movement of tiny particles (e.g., pollen) suspended in a liquid or a gas, or on the surface of a liquid).

Consider next the *box-fractal*: this is a square, in which the basic iteration is to divide it into 9 identical smaller squares and remove the middle squares from each side, leaving 5 of the original 9. It is readily seen that $a = 1/3$ and $N = 5$, so

$$D = \frac{\log 5}{\log 3} \approx 1.46497.$$

The *Sierpinski triangle* is any triangle for which the basic iteration is to join the midpoints of the sides with line segments and remove the middle triangle. Now $a = 1/2$ and $N = 3$, so

$$D = \frac{\log 3}{\log 2} \approx 1.58496.$$

These fractals penetrate increasingly more into the second dimension. We will mention two more at this juncture: the *Menger sponge* and *Cantor dust*. For the former, we do in three dimensions what was done in two for the box fractal. Divide a cube into 27 identical cubes and "push out" the middle ones in each face (and the central one). Now it follows that $a = 1/3$ and $N = 20$ (seven smaller cubes having been removed in the basic iterative step), so in the limit of the requisite infinite number of iterations

$$D = \frac{\log 20}{\log 3} \approx 2.72683,$$

(intruding well into the third dimension). Now for *Cantor dust*: what is that? Take any line segment and remove the middle third; this is the basic iteration. In the limit of an infinity of such iterations for which $a = 1/3$ and $N = 2$, it follows that

$$D = \frac{\log 2}{\log 3} \approx 0.63093,$$

which is obviously less than one. Quite amazing.

Appendix 10

RANDOM WALKS AND THE DIFFUSION EQUATION

Consider the motion of non-interacting "point particles" along the x-axis only, starting at time $t = 0$ and $x = 0$ and executing a *random walk*. The particles can be whatever we wish them to be: molecules, pollen, cars, inebriated people, white-tailed deer, rabbits, viruses, or anything else as long as there are sufficiently many for us to be able to assume a continuous distribution, and yet not so dense that they interact and interfere with one another (though this may seem rather like "having our cake and eating it"!). In its simplest description the particle motion is subject to the following constraints or "rules":

1. Every τ seconds, each particle moves to the left or the right, moving at speed v, a distance $\delta = \pm v\tau$. We consider all these parameters to be constants, but in reality they will depend on the size of the particle and the medium in which it moves.

2. The probability of a particle moving to the left is ½, and that of moving to the right is also ½. Thus they have no "memory" of previous steps (just like the toss of a fair coin); successive steps are statistically independent, so the walk is not biased.

3. As mentioned above, each particle moves independently of the others (valid if the density of particles in the medium is sufficiently dilute).

Some consequences of these assumptions are that (i) the particles go nowhere on the average and (ii) that their root-mean-square displacement is proportional, not to the time elapsed, but to the *square root* of the time elapsed. Let's see why this is so. With N particles, suppose that the position of the ith particle after the nth step is denoted by $x_i(n)$. From rule 1 it follows that $x_i(n) - x_i(n-1) = \pm\delta$. From rule 2, the $+$ sign will apply to about half the particles and the − sign will apply to the other half if N is sufficiently large; in practice this will be the case for molecules. Then the mean displacement of the particles after the nth step is

$$\langle x(n)\rangle = \frac{1}{N}\sum_{i=1}^{N} x_i(n) = \frac{1}{N}\sum_{i=1}^{N} [x_i(n-1)\pm\delta] = \frac{1}{N}\sum_{i=1}^{N} x_i(n-1) = \langle x(n-1)\rangle.$$

$$(A10.1)$$

That is, the mean position does not change from step to step, and since they all started at the origin, the mean position is still the origin! This means that the spread of the particles is symmetric about the origin.

Let's now consider the mean-square displacement $\langle x^2(n)\rangle$. It is clear that

$$x_i^2(n) = x_i^2(n-1) \pm 2\delta x_i(n-1) + \delta^2$$

so that

$$\langle x^2(n)\rangle = \frac{1}{N}\sum_{i=1}^{N} x_i^2(n) = \frac{1}{N}\sum_{i=1}^{N} [x_i^2(n-1) \pm 2\delta x_i(n-1) + \delta^2]$$

$$= \langle x^2(n-1)\rangle + \delta^2$$

$$(A10.2)$$

Since $x_i(0) = 0$ for all particles i,

$$\langle x^2(0)\rangle = 0, \text{ and hence } \langle x^2(1)\rangle = \delta^2, \langle x^2(2)\rangle = 2\delta^2, \dots, \text{ and } \langle x^2(n)\rangle = n\delta^2.$$

Thus the mean-square displacement increases as the step number n, and the root-mean-square displacement increases as the *square root* of the step number n. But the particles execute n steps in a time $t = n\tau$ so n is proportional to t. Thus the spreading is proportional to $\sqrt{t}$. And as we have shown in Chapter 19 for the case of a time-independent smokestack plume or a line of traffic, the spreading is proportional, in the same fashion, to the square root of distance from the source.

Incidentally, there is a very nice application of these ideas, extended to three dimensions, in the book by Ehrlich (1993). In an essay entitled "How slow can light go?" he points out that, barring exotic places like neutron stars and black holes, it's probably in the center of stars that light travels most slowly. This is because the denser the medium is, the slower is the speed of light.

In fact, it takes light generated in the central core of our star, the sun, about 100,000 years to reach the surface! (Do not confuse this with the 8.3 minutes it takes for light to reach Earth from the sun's surface.) The reason for this long time is that the sun is extremely opaque, especially in its deep interior. This means that light travels only a short distance (the *path length*) before being absorbed and re-emitted (usually in a different direction). The path of a typical photon of light traces out a *random walk* in three dimensions, much like the two-dimensional path of a highly inebriated person walking away from a lamppost. After a random walk consisting of N steps of length (say) 1 ft each, it would be very unlikely for the person to end up N feet away from the lamppost. In fact the average distance staggered away from the lamppost turns out to be $\sqrt{N}$ feet.

In the sun, the corresponding path length is about 1 mm, not 1 foot. This is the average distance light travels before being absorbed, based on estimates of the density and temperature inside the sun. The radius of the sun is about 700,000 km, or 7×10^5 km, or approximately 10^{12} mm, rounding up for simplicity to the nearest power of ten. Now watch carefully: there's nothing up my sleeve. We require this distance, the radius of the sun, to be the value of $\sqrt{N}$, the distance from the center traveled by our randomly "walking" photon. This means that $N \approx (10^{12})^2 = 10^{24}$, and 10^{24} mm steps is 10^{18} km. Now 1 light year is about 10^{13} km (check it for yourself!), so $N \approx 10^5$ light years, which by definition takes 10^5 or 100,000 years! Good gracious me!

THE DIFFUSION EQUATION

Now we formulate the one-dimensional diffusion equation in a very similar fashion. Basically, we wish to be able to describe the "density" of a distribution of "particles" along a straight line (the x-direction) as a function of time (t). $N(x,t)$ will denote the density of particles (number per unit volume) at location x and time t. How many particles at time t will move across unit area perpendicular to the x-direction from x to $x + \delta x$? By the time $t + \tau$ (i.e., during the next time step), half of the particles will have moved to the location $x + \delta$ and half of those located at $x + \delta$ will have moved to x (see Figure A10.1). This means that the net number moving from x to $x + \delta x$ is $[N(x,t) - N(x + \delta,t)]/2$. The total number of these particles per unit time and per unit area is called the *net flux*. We'll call this F_x and rearrange it as

$$F_x = \frac{[N(x,t) - N(x + \delta,t)]}{2A\tau} = -\frac{\delta^2}{2\tau}\frac{1}{\delta}\left[\frac{N(x + \delta,t) - N(x,t)}{\delta A}\right]. \quad \text{(A10.3)}$$

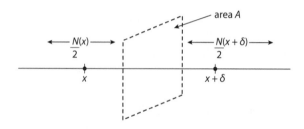

Figure A10.1. Schematic for the flux argument leading to equation (A10.4).

The quantity $\delta^2/2\tau$ has dimensions of $(\text{distance})^2/(\text{time})$, and this will be called the *diffusion coefficient D*. The first quotient in square brackets is the number of particles per unit volume at $x + \delta$ and time t. This is just the concentration, denoted by $C(x + \delta, t)$. Similarly, the second term is $C(x, t)$. This means that we can rewrite the net flux as

$$F_x = -\frac{D}{\delta}[C(x + \delta, t) - C(x, t)].\qquad (A10.4)$$

At this point we are in a position to carry out a familiar procedure: taking the limit of this quotient as $\delta \to 0$. If this limit exists, we can write

$$F_x = -D\frac{\partial C}{\partial x}.\qquad (A10.5)$$

Physically this means that the net flux is proportional to the concentration gradient, and it moves in the opposite direction. We can think of it this by imagining instead that C is temperature; the flow of heat will be from the higher temperature region toward the lower one—in a direction opposite to the gradient of the temperature. If on the other hand C is the concentration of sugar in my tea, the flow of sugar molecules is toward regions of lower concentration.

Let's take this one stage farther and consider a little slab of thickness δ and area A perpendicular to the x-axis (see Figure A10.2). In a time τ, the number of particles entering from the left is $F_x(x)A\tau$, while $F_x(x + \delta)A\tau$ leave from the right (assuming no particles are created or destroyed). This means that the number of particles per unit volume in the slab must increase at a rate given by the expression

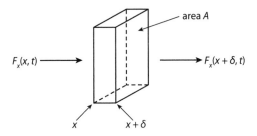

Figure A10.2. Schematic for the flux argument leading to equation (A10.6).

$$\frac{C(t+\tau)-C(t)}{\tau} = \frac{A\tau\left[F_x(x,t+\tau)-F_x(x+\delta,t+\tau)\right]}{A\delta}$$
$$= -\frac{\left[F_x(x+\delta,t+\tau)-F_x(x,t+\tau)\right]}{\delta}. \qquad (A10.6)$$

In the limit $\tau \to 0$, $\delta \to 0$ we obtain

$$\frac{\partial C}{\partial t} = -\frac{\partial F_x}{\partial x} = D\frac{\partial^2 C}{\partial x^2}. \qquad (A10.7)$$

There is another mechanism that must be included in any realistic discussion of pollution: wind. As with the discussion in Chapter 19 we shall consider the effects of a wind with constant speed U in the x-direction only (even when a higher-dimensional diffusion equation is used). The rate at which particles enter the slab per unit volume is approximately $UC(x,t)A$, so the wind's contribution to the right-hand side of the diffusion equation is approximately

$$UA\frac{\left[C(x,t)-C(x+\delta,t)\right]}{A\delta};$$

And so in the usual limit, equation (A10.7) is generalized slightly to become

$$\frac{\partial C}{\partial t} = D\frac{\partial^2 C}{\partial x^2} - U\frac{\partial C}{\partial x}. \qquad (A10.8)$$

RAINBOW/HALO DETAILS

My heart leaps up when I behold a rainbow in the sky
—*William Wordsworth*

In Chapter 22 a contrast was drawn between some meteorological optical effects—rainbows and some ice crystal halos—observed during the day, from those potentially observable from nearby light sources at night. This Appendix summarizes some of the salient features of rainbows and one of the common ice crystal halos.

So what is a rainbow, and what causes it? A rainbow is sunlight, displaced by reflection and dispersed by refraction in raindrops, seen by an observer with his or her back to the sun. The primary rainbow, which is the lower and brighter of two that may be seen, is formed from two refractions and one reflection in myriads of raindrops (see Figure A11.1). It can be seen and photographed, but it is not located at a specific place, only in a particular set of directions. Obviously the raindrops causing it are located in a specific region in front of the observer. The path for the secondary rainbow is similar, but involves one more internal reflection. In principle, an unlimited number of higher-order rainbows exist from a single drop, but light loss at each reflection limits the number of visible rainbows to two. Claims have been made concerning observations of a tertiary bow (and even a quaternary bow), however, such a bow would occur around the sun, and be very difficult to observe, quite apart from its intrinsic faintness. Nevertheless, in 2011, significant photographic evidence for such bows was published in a reputable scientific journal (see Großmann et al. (2011) and Theusner (2011)); it will be exciting to see what further research is carried out in this area.

Returning to the primary bow, note that while each drop produces its individual primary rainbow, what is seen by an observer is the cumulative set of images from myriads of drops, some contributing to the red region of the bow, others to the orange, yellow, green, and so forth. Although each drop is falling, there are numerous drops to replace each one as it falls through a particular location, and so the rainbow, for the period that it lasts, is for each observer effectively a continuum of colors produced by a near-continuum of drops.

Let's start with an examination of the basic geometry for a light ray entering a spherical droplet. From Figure A11.1 note that after two refractions and one reflection the light ray shown contributing to the rainbow has undergone a total deviation of $D(i)$ radians, where

$$D(i) = \pi + 2i - 4r \qquad (A11.1)$$

in terms of the angles of incidence (i) and refraction (r), respectively. The latter is a function of the former, this relationship being expressed in terms of Snell's law of refraction,

$$\sin i = n \sin r, \qquad (A11.2)$$

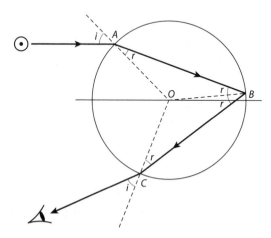

Figure A11.1. Ray path inside a spherical drop in the formation of a primary rainbow from sunlight.

where n is the *relative index of refraction* (of water, in this case). This relative index is defined as

$$n = \frac{\text{speed of light in medium I (air)}}{\text{speed of light in medium II (water)}} > 1.$$

Since the speed of light in air is almost that "in vacuo," we will refer to n for simplicity as the refractive index; its generic value for water is $n \approx 4/3$, but it does depend slightly on wavelength (this is the phenomenon of dispersion, and without it we would only have bright "white bows"!)

How does the deviation angle D vary with the angle of incidence i? Note that for rays that enter the sphere along the axis of symmetry and are reflected from the back surface, $D(0) = \pi$. The classic "rainbow problem" is to find whether there are maxima or minima in this deviation angle as a function of the angle i. On differentiating equation (A11.1) with respect to i, and using equation (A11.2), it is found that $D'(i) = 0$ when

$$i = i_c \equiv \arccos\left(\frac{n^2 - 1}{3}\right)^{1/2} \approx 59.4° \qquad (\text{A11.3})$$

for $n = 4/3$, on converting to degrees of arc. This means that D is stationary at $i = i_c$. Physically, this corresponds to a concentration of deviated rays in a small angular region about

$$D(i_c) = \pi + 2i_c - 4\arcsin\left(\frac{\sin i_c}{n}\right) \approx 138°. \qquad (\text{A11.4})$$

It can be shown that $D''(i_c) > 0$, so that the deflection $D(i_c)$ is a minimum (in fact it is a global minimum in $(0, \pi/2)$ because $D''(i_c) > 0$). This minimum angle of deviation is often referred to as the *rainbow angle*. Its supplement $180° - D(i_c) \approx 42°$ is the semi-angle of the rainbow "cone" formed with apex at the observer's eye, the axis being the line joining the eye to the shadow of her head (the antisolar point).

A comment about the color dispersion in a rainbow is in order. While it is not standardized to the satisfaction of everyone, the visible part of the electromagnetic spectrum extends from the "red" end (700–647 nm, or 0.700μ– 0.647μ, etc.) to the "violet" end (424–400 nm), a nanometer (nm) being 10^{-9} m. For red light of wavelength $\lambda \approx 656$ nm, the refractive index $n \approx 1.3318$, whereas for violet light of wavelength $\lambda \approx 405$ nm, the refractive index

$n \approx 1.3435$, a slight but very significant difference! All that has to be done is the calculation of i_c and $D(i_c)$ for these two extremes of the visible spectrum, and the difference computed. Then voilà! We have the angular width of the primary rainbow. In fact since $D(i_c) \approx 137.8°$ and $139.4°$ for the red and violet ends, respectively, the angular width is $\Delta D \approx 1.6°$, or about three full moon angular widths.

There is also a secondary rainbow that is frequently visible along with the primary bow as a result of an extra reflection inside the raindrops. As a result it is the fainter of the two, slightly wider, and about 9 degrees "higher" in the sky than the primary bow.

What about the common halo mentioned above? Have you ever noticed a circular ring around the sun (or, for that matter, the moon) when the sky is clear except for wispy thin cirrus clouds in the vicinity of the sun? I read on a now-defunct website that such beautiful displays, known as ice crystal halos, can be seen on average twice a week in Europe and parts of the United States, and certainly my own experience is not terribly different, though I would estimate that I notice them about three times a month on average. The "radius" of a ring around the sun or moon is naturally expressed in terms of degrees of arc, subtended at the observer's eye by the apparent radius, just as for the rainbow. The most frequently visible one is the 22° circular halo, followed by parhelia (or *sundogs*, colored "splotches" of light on one side or other of the sun, and commonly both). In my own rather limited experience, the sundogs are as common as the halo if not more so, at least as I walk to work in southeastern Virginia. These are found at the same altitude as the sun, close to (but just beyond) the 22° halo. A convenient and literal "rule of thumb" is that the outstretched hand at arm's length subtends about the same angle, so that if one's thumb covers the sun, one's little finger extends to about the 22° halo (or the sundog). There are *many* other types of ice crystal displays; for more information consult the encyclopedic and highly regarded website by atmospheric optics expert Les Cowley [46].

Halos are formed when sunlight is refracted, reflected, or both from ice crystals in the upper atmosphere and enters the eye of a careful observer: be careful, because even the common types are easily missed (but never look directly at the sun, of course). As already noted, they are often produced when a thin uniform layer of cirrus or cirrostratus cloud covers large portions of the sky, especially in the vicinity of the sun. Surprisingly, perhaps, they may occur at any time of the year, even during high summer, because above an altitude of about 10 km it is always cold enough for ice crystals to form. In particularly cold climes, of course, such crystals can form at ground level (though we are not thinking here of snow crystals). Such halos can arise from *diamond dust*, which is essentially ground-level cloud composed of tiny ice crystals. It can form wherever the temperature is well below freezing. Of course, some types of halo (such as a *circumhorizontal arc*) are very latitude-dependent, and may therefore be seen only rarely in higher latitudes. Very many of the crystals producing halos are hexagonal prisms; some are thin flat plates while others are long columns, and sometimes the latter have bullet-like or pencil-like ends. A significant feature of all these crystals is that while any given type may have a range of sizes, the angles between the faces are the same. Although they do not possess perfect hexagonal symmetry, of course, they are sufficiently close to this that simple geometry based on such idealized forms suffices to describe the many different arcs and halos that are seen. The halos that result from cirrus cloud crystals depend on two major factors: their shape and their orientation as they fall. Their shape is determined to a great extent by their history, that is, the temperature of the regions through which they drift as they are drawn down by gravity and buffeted around by winds and convection currents.

In order to explain the reasons for the various angles, for example, 22°, it is necessary to examine the crystal geometry in more detail. It is clear from Figure A11.2 that hexagonal ice crystals can be thought of as presenting "prism angles" of 60°, 90°, or 120° to rays entering them in the planes indicated, depending on the orientation of the crystal. There are both similarities

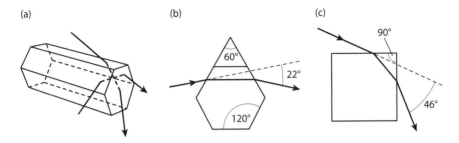

Figure A11.2. Ray paths through hexagonal ice crystal prisms for both the 22° and the rarer 46° halos.

and differences with the formation of rainbows in raindrops, and noting that a sphere is (infinitely) more symmetrical than a regular hexagonal plate, a result from optics that is extremely important for such halos is *that the deviation angle for light refracted through a prism is a minimum for symmetric ray paths.*

Recall that a rainbow arises when the angle through which a ray is deviated on passing through the raindrop is an extremum. The above result tells us that in the case of prisms, the corresponding extremum—also a minimum, in fact—occurs when the ray path through the crystal is symmetric. A second result from elementary optics defines the magnitude of this minimum deviation angle (D_m)—the location of the halo relative to the line joining the sun and the observer—in terms of the refractive index (of ice here) and the apex angle of the prism.

We can use these results to explain the occurrence of the 22° and the less common 46° halos. As shown in Figure A11.2 (b,c), there are three prism angles in a hexagonal ice crystal prism: 60° (light entering side 1 and exiting side 3); 90° (light entering a top or bottom face and exiting through a side) and 120° (light entering side 1 and being totally internally reflected by side 2). The first two of these create color by dispersion of sunlight; the last contributes nothing directly to a halo, at least of interest to us here. The refractive index of ice for yellow light is $n \approx 1.31$. For an apex angle of 60°, $D_m \approx 22°$, and for an apex angle of 90°, $D_m \approx 46°$. As in the case of the rainbow, all possible deviations are present in reality, but it is the "clustering" of deviated rays near the minimum that provides the observed intensity in the halos (but unlike the case of the rainbow, no reflection contributes to their formation in these two cases).

Appendix 12

THE EARTH AS VACUUM CLEANER?

How effective is the Earth at clearing a path through "space matter" in its vicinity? This is of interest, of course, with regard to possible close encounters with asteroids, as discussed in Chapter 24. In order to get a handle on this problem, one major task is to determine how effective the Earth is at capturing "errant" asteroids. In order to do so, we will use some elementary physical principles and properties of conic sections.

Because of the gravitational attraction of the Earth (or any other sufficiently massive body), it can in principle "pull in" objects that are not traveling directly toward a head-on collision. This results in a *capture cross section* (CCS) that is at least as large as its *geometric cross section* (GCS). It is clear that this will depend (in particular) on the speed of the object relative to the Earth; something "whizzing by" our planet at high speed is less likely to be captured than a much slower object on the same path.

We can reduce this to a simple "two-body" problem by ignoring, for now, the rest of universe (as I so often do). The approach here is based on the article by Tatum (1997) [47]. Essentially, we consider the trajectory of the asteroid to be such that the gravitational attraction of the Earth is the dominant mechanism in this encounter. This is reasonable since the force of attraction of the Earth is ten times that of the sun at a distance of about 52,000 miles, or 13 Earth radii from its center, and about 1700 times that of the moon at this same distance. Under this assumption we can regard the trajectory of the asteroid as a hyperbolic one about a stationary Earth. Of course, both objects are in elliptical orbits around the sun, but in this "geocentric model" a hyperbolic orbit around the Earth is perfectly adequate for a "back of the envelope" calculation such as this.

The notation is as follows (see Figure A12.1). The impact parameter B (no pun intended) is the closest distance of approach to Earth's center that the asteroid would have it were on a straight line path, that is, one unaffected by

Earth's gravitational attraction (or more accurately, by their mutual gravitational attraction). Its initial speed (at "infinity") is v_0 and the equation of its hyperbolic path is

$$\frac{x^2}{a^2} - \frac{y^2}{b^2} = 1, \qquad (A12.1)$$

a being the x-intercept and b the semi-transverse axis of the conjugate hyperbola.

Exercise: Using Figure A12.1, show that the impact parameter

$$B = \frac{b}{a}(a^2 + b^2)^{1/2} \approx b \text{ if } b \ll a.$$

The closest distance of approach on the hyperbolic orbit is called the *perigee distance p*, and if R is the radius of the Earth, the asteroid will collide with the Earth if $p < R$ (the case $p = R$ corresponds to grazing incidence). The GCS of the Earth is πR^2 and the potential CCS is defined to be πB^2. How do these two areas compare as a function of the initial speed v_0?

First we will need to calculate the potential energy $\Phi(\bar{r})$ of the asteroid under the action of the Earth's gravitational field, where $\bar{r}$ is the position vector

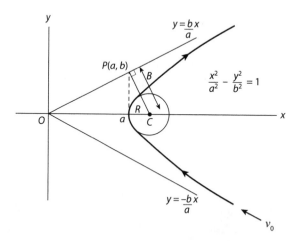

Figure A12.1. The asteroid's hyperbolic orbit drawn for "grazing incidence" with the Earth. The impact parameter **B** is the perpendicular distance of the asymptote from the Earth's center of attraction at **C**. The radius of the Earth is **R** and the speed of the asteroid "at infinity" is v_0.

of the asteroid relative to the Earth. The gravitational force $\bar{F}$, described by the famous inverse-square law, is directed along the radius vector $\bar{r}$ joining the two bodies, and for a spherically symmetric force Φ is given by

$$\Phi(r) = \Phi(|\bar{r}|) = -\int_r^\infty \bar{F} \cdot d\bar{\xi} = -\int_r^\infty \frac{Gm_a M_e}{\xi^2} d\xi = -\frac{Gm_a M_e}{\xi}, \quad (A12.2)$$

where m_a is the mass of the asteroid, M_e is the mass of the Earth, and G is the gravitational constant. At the perigee distance we suppose the asteroid speed relative to the Earth to be v_p, and we apply the principle of the conservation of energy to the asteroid, namely, that the

(kinetic energy + potential energy) at "infinity"
= (kinetic energy + potential energy) at perigee,

that is,

$$\frac{1}{2} m_a v_0^2 = \frac{1}{2} m_a v_p^2 - \frac{Gm_a M_e}{p},$$

or, simplifying,

$$v_0^2 = v_p^2 - \frac{2GM_e}{p}. \quad (A12.3)$$

Another fundamental principle is the conservation of angular momentum,

(mass × velocity × impact parameter) at "infinity"
= (mass × velocity × impact parameter) at perigee,

that is,

$$v_0 B = v_p p. \quad (A12.4)$$

Elimination of v_p from these two equations results in

$$\frac{B^2}{R^2} = 1 + \frac{v_e^2}{v_0^2}, \quad (A12.5)$$

In this equation $v_e = (2GM_e/R)^{1/2}$ is the escape velocity from the surface of the Earth. To see this we equate the "escape" kinetic energy with the potential energy of a particle of unit mass at infinity. Thus

$$\frac{1}{2}v_e^2 = \frac{GM_e}{R} \tag{A12.6}$$

from which the result follows. Incidentally, $v_e \approx 11.2$ km/s. Equation (A12.5) gives that the ratio of capture-to-geometric cross sections has a lower bound of one, as would be expected. It differs significantly from this value only when $v_0 < v_e$; thus for $v_0 = 0.7v_e$, $B^2/R^2 \approx 3$.

As pointed out by Tatum [47], there *is* a class of asteroids that approach our planet quite closely, and with small relative speeds, so the CCS/GCS ratio for these may be quite large. One such object is 1991 VG, which at that time was, at eight meters, the smallest astronomical object ever discovered (in orbit). The semi-major axis of its orbit is 1.04 astronomical units ($\approx 1.50 \times 10^8$ km; the mean Earth-sun distance is approximately 1 A.U.). The eccentricity of its orbit is 0.067 (Earth's is 0.0167), which brings it inside the Earth's orbit for part of the time. With this in mind, let the Earth move again (Oh, the power of mathematics!), in a circular orbit for simplicity, and imagine such an asteroid in a nearby circular orbit. Obviously we expect that these orbits will not be concentric in practice, but they may be close enough for a significant portion of the orbit to make this simplification useful. If the distance apart of the two orbits is less than the radius of the CCS, then the object can be captured. The closer the orbit is to that of the Earth, the smaller the relative speed and the larger the CCS becomes.

Suppose that the speed of the Earth in its orbit is V with orbit radius a, and the corresponding speed and orbit radius for the asteroid are $V + \delta V$ and $a + \delta a$, respectively. Now Kepler's third law of planetary motion states that the square of the orbital period P of a planet is directly proportional to the cube of the semi-major axis of its orbit (here a), or in terms of a constant of proportionality K, $P^2 = Ka^3$. For a circular orbit, $P = 2\pi a/V_e$ so that on eliminating P, Kepler's law reduces to

$$V^2 = \frac{4\pi^2}{Ka} \equiv \frac{\alpha}{a}.$$

Using differentials it follows that the relative difference in speed between the two objects is

$$\frac{\delta V}{V} = -\frac{\alpha \delta a}{2V^2 a^2} = -\frac{\delta a}{2a}. \tag{A12.7}$$

From the earlier discussion we can identify δV, the relative difference in speeds, as v_0 and the difference in orbital radii, δa, as the impact parameter b. Using equation (A12.7) in (A12.5) we obtain (in the current notation)

$$(\delta a)^4 - R^2(\delta a)^2 - \frac{8GMRa^2}{V^2} = 0.$$

The positive real root of this biquadratic equation is

$$\delta a = \left\{ \frac{R^2 + [R^4 + 32GMRa^2/V^2]^{1/2}}{2} \right\}^{1/2}. \qquad \text{(A12.8)}$$

In performing these calculations I found it simplest to write the term

$$\frac{32GMRa^2}{V^2} \quad \text{as} \quad 16\left(\frac{2GM}{R}\right)R^2\left(\frac{T}{2\pi}\right)^2$$

since $a/V = T/2\pi$, where T is the period of the Earth's orbit in seconds. We have already calculated the quantity $2GM/R$ from equation (A12.6). The result is $\delta a \approx 8.45 \times 10^5$ km ($\delta a \approx 5.63 \times 10^{-3}$ A.U.). In terms of the Earth's equatorial radius $R \approx 6.38 \times 10^3$ km, this is about 132 Earth radii, or approximately 2.2 times the Earth-moon semi-major axis.

From equation (A12.7) we find that the corresponding difference in orbital speed $|\delta V| \approx 0.085$ km/s.

Exercise: Show from equations (A12.2) and (A12.5) that, for a grazing collision,

$$|v_p| = \frac{VB^2}{2aR}.$$

This shows that the perigee speed is quite sensitive to the impact parameter, varying as its square. Using this result we calculate the impact speed at a grazing collision to be ≈ 11.1 km/s, just a tiny bit less than the escape speed from the Earth. Since we have shown that the Earth sweeps out a toroidal volume of radius approximately 0.056 A.U. in its path around the sun, it can be thought of as a giant vacuum cleaner with CCS $(132)^2 \approx 1.74 \times 10^4$ times that of the Earth's GCS. Therefore any object moving in a circular orbit with radius between 0.9944 A.U. and 1.0056 A.U. will collide with (be captured by) the Earth. That's some vacuum cleaner!

ANNOTATED REFERENCES AND NOTES

NOTES

[1] Batty, M. and Longley, P. (1994). *Fractal Cities*. Academic Press, San Diego, CA.

[2] Polya, G. (1954). *Mathematics and Plausible Reasoning. I. Induction and Analogy in Mathematics*. Princeton University Press, Princeton, NJ.

[3] Hern, W.M. (2008). "Urban malignancy: similarity in the fractal dimensions of urban morphology and malignant neoplasms." *International Journal of Anthropology* 23, 1–19.

[4] See Adam (2006) for a slightly different version of the Princess Dido story.

[5] Barshinger, R. (1992). "How not to land at Lake Tahoe!" *American Mathematical Monthly* 99, 453–455.

[6] Taxi rides, squirrels, and light bulbs: I am grateful to my friend, colleague, and co-author Larry Weinstein, an NYC boy, for his insights into city life; I have drawn on his expertise for these questions.

[7] Weinstein, L. and Adam, J.A. (2008). *Guesstimation: Solving the World's Problems on the Back of a Cocktail Napkin*. Princeton University Press, Princeton, NJ.

[8] Eastaway, R. and Wyndham, J. (1999). *Why do Buses Come in Threes? The Hidden Mathematics of Everyday Life*. Wiley, London.

[9] Skyscrapers and harmonic motion: see http://www.cpo.com/pdf/tpst_pfc%20ch19%20connections.pdf.

[10] Burghes, D., Galbraith, P., Price, N. and Sherlock, A. (1996). *Mathematical Modelling*. Prentice-Hall, Hempstead, Hertfordshire.

[11] Barnes, G. (1990). "Food, eating, and mathematical scaling." *Physics Teacher* 28, 614–615.

[12] Burton, R.F. (1998). *Biology by Numbers*. Cambridge University Press, Cambridge.

[13] Edwards, D. and Hamson, M. (1989). *Guide to Mathematical Modelling*. CRC Press, Boca Raton, FL.

[14] Lowson, M.V. (2004). "Idealised models for public transport systems." *International Journal of Transport Management* 2, 135–147.

[15] Lowson, M.V. (1999). "Personal public transport." *Proceedings of the Institution of Civil Engineers: Transportation* 135, 139–151.

[16] Mahajan, S. (2010). *Street-Fighting Mathematics: The Art of Educated Guessing and Opportunistic Problem Solving,* MIT Press, Cambridge, MA.

[17] Seifert, H.S. (1962). "The stop-light dilemma." *American Journal of Physics* 30, 216–218.

[18] Chapman, S. (1942). "Should one stop or turn in order to avoid an automobile collision?" *American Journal of Physics* 10, 22–27.

[19] Ashdon, W.D. (1966). *The Theory of Road Traffic Flow*. Methuen, London.

[20] Smeed, R.J. (1961). *The Traffic Problem in Towns*. Manchester Statistical Society, Manchester, England .

[21] Smeed, R.J. (1963). "The road space required for traffic in towns." *Town Planning Review* 33, 270–292.

[22] Smeed, R.J. (1963). "Road development in urban areas." *Journal of the Institute of Highway Engineers* 10, 5–26.

[23] Smeed, R.J. (1965). "A theoretical model for commuter traffic in towns." *Journal of the Institute of Mathematics and its Applications* 1, 208–225.

[24] Smeed, R.J. (1967). "The road capacity of city centers." *Highway Research Record* 169, 22–29.

[25] Smeed, R.J. (1968). "Traffic studies and urban congestion." *Journal of Transport Economics and Policy* 2, 33–70.

[26] Smeed, R.J. (1970). "The capacity of urban road networks." *Proceedings of the Australian Road Research Board* 5, 10–28.

[27] Smeed, R.J. (1977). "Traffic in a linear town." *Proceedings of the 7th International Symposium on Transportation and Traffic Theory.*

[28] von Foerster, H., Mora, P.M., and Amiot, L.W. (1960). "Doomsday: Friday 13 November, A.D. 2026." *Science* 132, 1291–1295.

[29] Johansen, A. and Sornette, D. (2001). "Finite-time singularity in the dynamics of the world population, economic and financial indices." *Physica A* 294, 465–502.

[30] Bracken, A.J., and Tuckwell, H.C. (1992). "Simple mathematical models for urban growth." *Proceedings of the Royal Society of London A* 438, 171–181.

[31] Zipf, G.K. (1949). *Human Behavior and the Principle of Least Effort*. Harvard University Press, Cambridge, MA.

[32] Bettencourt, L.M.A., Helbing, J.L.D., Kühnert, C. and West, G.B. (2007). "Growth, innovation, scaling, and the pace of life in cities." *Proceedings of the National Academy of Sciences of the USA* 104, 7301–7306; also Bettencourt, L.M.A., Lobo, J. and West, G.B. (2008). "Why are large cities faster? Universal scaling and self-similarity in urban organization and dynamics." *European Physical Journal B* 63, 285–293.

[33] Ide, K., and Sornette, D. (2002). "Oscillatory finite-time singularities in finance, population and rupture." *Physica A* 307, 63–106.

[34] Bosanquet, C.H., and Pearson, J.L. (1936). "The spread of smoke and gases from chimneys." *Transactions of the Faraday Society* 32, 1249–1263.

[35] http://www.ias.ac.in/initiat/sci_ed/resources/chemistry/LightScat.pdf.

[36] Lynch, D., and Livingston, W. (2010). *Color and Light in Nature.* 2nd ed., reprint. Thule Scientific, Topanga, CA; p.26.

[37] http://antwrp.gsfc.nasa.gov/apod/ap040913.html.

[38] Minnaert, M. (1954). *The Nature of Light and Colour in the Open Air.* Dover, New York. (German edition, 1993, *Light and Color in the Outdoors*, Springer-Verlag, Berlin.)

[39] See website by Christian Fenn, "Rainbows in diverging light." http://www.meteoros.de/rainbow/rbdive_1.htm.

[40] http://quirkynyc.com/2010/05/ahoy-lighthouses-litter-new-york-city/.

[41] http://www2.jpl.nasa.gov/sl9/.

[42] Banks, R.B. (1998). *Towing Icebergs, Falling Dominoes, and Other Adventures in Applied Mathematics.* Princeton University Press, Princeton, NJ.

[43] http://www.citymayors.com/statistics/largest-cities-area-125.html; see also http://www.worldatlas.com/citypops.htm.

[44] Taxicab geometry: http://en.wikipedia.org/wiki/Taxicab_geometry.

[45] http://www.sci.csuhayward.edu/statistics/Resources/Essays/BinPois.htm and https://mywebspace.wisc.edu/adjacob1/soc_361/handouts/poisson%20handout.pdf

[46] See http://www.atoptics.co.uk/.

[47] Tatum, J.B. (1997). "The capture cross-section of earth for errant asteroids." *Journal of the Royal Astronomical Society of Canada* 91, 276–278.

OTHER RELEVANT LINKS

Chapter 15: Adapted from the question posed by Nathan Keyfitz, "How many people have lived on the Earth?" See also http://www.prb.org/Articles/2002/HowMany PeopleHaveEverLivedonEarth.aspx.

Chapter 18: See also http://brenocon.com/blog/2009/05/zipfs-law-and-world-city -populations/ (an excellent link) and http://opinionator.blogs.nytimes.com/2009/ 05/19/math-and-the-city/ (an informative article by Steven Strogatz).

Chapter 25: http://www.solarviews.com/eng/comet/appendc.htm (by Calvin Hamilton).

GENERAL REFERENCES

Adam, J.A. (1988). "Complementary levels of description in applied mathematics—III: equilibrium models of cities." *Mathematical and Computer Modelling* 10, 321–339.

Adam, J.A. (2006). *Mathematics in Nature: Modeling Patterns in the Natural World.* Paperback ed. Princeton University Press, Princeton, NJ.

Adam, J.A. (2011). *A Mathematical Nature Walk.* Paperback ed. Princeton University Press, Princeton, NJ.

Adams, G.U. (1987). "Smeed's law: some further thoughts." *Traffic Engineering and Control* 28, 70–73.

Alvarez, L.W., Alvarez, W., Asaro, F., and Michel, H.V. (1980). "Extraterrestrial Cause for the Cretaceous-Tertiary Extinction." *Science* 208, 1095–1108.

Amson, J.C. (1972). "Equilibrium models of cities: 1. An axiomatic theory." *Environment and Planning* 4, 429–444.

Amson, J.C. (1973). "Equilibrium models of cities: 2. Single-species cities." *Environment and Planning* 5, 295–338.

Angel, S., and Hyman, G.M. (1970). "Urban velocity fields." *Environment and Planning* 2, 211–224.

Angel, S., and Hyman, G.M. (1976). *Urban Fields.* Pion, London.

Apsimon, H.G. (1958). "Note 2754. A repeated integral." *Mathematical Gazette* 42, 52.

Atkinson, B.W. (1968). "A Preliminary examination of the possible effects of London's urban area on the distribution of thunder rainfall 1951–1960." *Transactions of the Institute of British Geographers* 44, 97–108.

Batty, M. (1971). "Modelling cities as dynamic systems." *Nature* 231, 425–428.

Batty, M., Carvalho, R., Hudson-Smith, A., Milton, R., Smith, D., and Steadman, P. (2008). "Scaling and allometry in the building geometries of Greater London." *European Physical Journal B* 63, 303–314.

Billah, K.Y., and Scanlan, R.H. (1991). "Resonance, Tacoma Narrows bridge failure, and undergraduate physics textbooks." *American Journal of Physics* 59, 118–124.

Boas, R. (2002). Guest essay, "Kinematics of jogging." pp. 351–352 in M. J. Strauss, G. L. Bradley and K. J. Smith, *Calculus.* 3rd ed. Prentice-Hall, Englewood Cliffs, NJ.

Brakman, S., Garretsen, H., and van Marrewijk, C. (2009). *The New Introduction to Geographical Economics.* Cambridge University Press, Cambridge. 2nd ed.

Cohen, J.E. (1995). "Population growth and Earth's human carrying capacity." *Science* 269, 341–346.

Clayden, A.W. (1891). "On Brocken spectres in a London fog." *Quarterly Journal of the Royal Meteorological Society* 17, 209–216.

Crawford, F.S. (1988). "Rainbow dust." *American Journal of Physics* 56, 1006–1009.

Drew, D.R. (1968). *Traffic Flow Theory and Control.* McGraw-Hill, New York.

Einhorn, S.J. (1967). "Polar vs. rectangular road networks," *Operations Research* 35, 546–548.

Ehrlich, R. (1993). *The Cosmological Milkshake.* Rutgers University Press, New Brunswick, NJ.

Fairthorne, D.B. (1964). "The distances between random points in two concentric circles." *Biometrika* 51, 275–277.

Fairthorne, D.B. (1965). "The distance between pairs of points in towns of simple geometrical shape." *Proceedings of the Second International Symposium on Theory of Traffic flow* (J. Almond, ed.). OECD, Paris, 391–406.

Floor, C. (1980). "Rainbows and haloes in lighthouse beams." *Weather* 35, 203–208.

Floor, C. (1982). "Optic phenomena and optical illusions near lighthouses." *Meteorologische Zeitschrift* 32, 229–233.

Foster, J.H., and Pedersen, J.J. (1980). "On the reflective property of ellipses." *American Mathematical Monthly* 87, 294–297.

Garwood, F., and Tanner, J.C. (1958). "Note 2800. On note 2754: a repeated integral." *Mathematical Gazette* 52, 292–293.

Gerlough, D.L. (1955). *Use of Poisson Distribution in Highway Traffic.* ENO Foundation for Highway Traffic Control, Saugatuck, CT.

Giordano, F.R., Weir, M.D., and Fox, W.P. (2003). *A First Course in Mathematical Modeling.* 3rd ed. Thomson Brookes/Cole, Pacific Grove, CA.

Gislén, L., and Mattsson, J.O. (2003). "Observations and simulations of some divergent-light halos." *Applied Optics* 42, 4269–4279.

Gislén, L., and Mattsson, J.O. (2007). "Tabletop divergent-light halos." *Physics Education* 42, 579–584.

Großmann, M., Schmidt, E. and Haußmann, A. (2011). "Photographic evidence for the third-order rainbow." *Applied Optics* 50, F134–F141.

Haberman, R. (1977). *Mathematical Models: Mechanical Vibrations, Population Dynamics, and Traffic Flow,* Prentice-Hall, Englewood Cliffs, NJ.

Haight, F.A. (1964). "Some probability distributions associated with commuter travel in a homogeneous circular city." *Operations Research* 12, 964–975.

Harsch, J., and Walker, J.D. (1975). "Double rainbow and dark band in searchlight beam." *American Journal of Physics* 43, 453.

Herman, R., and Gardels, K. (1963). "Vehicular traffic flow." *Scientific American* 209, 35–43.

Hobbs, F.D., and Richardson, B.D. (1967). *Traffic Engineering,* vol. 2. Pergamon Press, Oxford.

Holroyd, E.M. (1969). "Polar and rectangular road networks for circular cities." *Transportation Science* 3, 86–88.

Hunt, J.C.R. (1971). "The effect of single buildings and structures." *Philosophical Transactions of the Royal Society of London A* 269, 457–467.

Ioannides, Y.N., and Overman, H.G. (2003). "Zipf's law for cities: an empirical examination." *Regional Science and Urban Economics* 33, 127–137.

Ishikawa, H. (1980). "A new model for the population density distribution in an isolated city." *Geographical Analysis* 12, 223–235.

Jacobs, J. (1984). *Cities and the Wealth of Nations.* Vintage, New York.

Jones, T.R., and Potts, R.B. (1962). "The measurement of acceleration noise—a traffic parameter." *Operations Research* 10, 745–763.

Kerensky, O.A.(1971). "Bridges and other large structures." *Philosophical Transactions of the Royal Society of London A* 269, 343– 351.

Keyfitz, N. (1976). *Applied Mathematical Demography.* Wiley, New York.

Kilminster, C.W. (1976) "Population in cities." *Mathematical Gazette* 60, 11–24.

Krause, E.F. (1986). *Taxicab Geometry: An Adventure in Non-Euclidean Geometry.* Dover, New York.

Lew, J.S., Frauenthal, J.C., and Keyfitz, N. (1978). "On the average distances in a circular disc." *SIAM Review* 20, 584–592.

Lowry, W.P. (1967). "The climate of cities." *Scientific American* 217, 15–23.

Mair, W.A., and Maull, D.J. (1971). "Aerodynamic behaviour of bodies in the wakes of other Bodies." *Philosophical Transactions of the Royal Society of London A* 269, 425–437.

Makse, H.A., Andrade, J.S., Jr., Batty, M., Havlin S. and Stanley, H.E. (1998). "Modeling urban growth patterns with correlated percolation." *Physical Review E* 58, 7054–7062.

Makse, H.A., Havlin, S., and Stanley, H.E. (2002). "Modelling urban growth patterns." *Nature* 377, 608–612.

Mallmann, A.J., Hock, J.L., and Greenler, R.G. (1998). "Comparison of sun pillars with light pillars from nearby light sources." *Applied Optics* 37, 1441–1449.

Mattsson, J.O., Nordbeck, S., and Rystedt, B. (1971). "Dewbows and fogbows in divergent light," No. 11 of *Lund Studies in Geography, Series C,* Lund University, Lund, Sweden.

Mattsson, J.O. (1973). "'Sun-sun' and light-pillars of street lamps."*Weather* 28, 66–68.

Mattsson, J.O. (1974). "Experiments on horizontal haloes in divergent light," *Weather* 29, 148–150.

Mattsson, J.O. (1978). "Experiments on the horizontal circle in divergent light." *Meteorologische Zeitschrift* 28, 123–125.

Mattsson, J.O. (1998). "Concerning haloes, rainbows and dewbows in divergent light." *Weather* 53, 176–181.

Mattsson, J.O., Bärring, L., and Almqvist, E. (2000). "Experimenting with Minnaert's cigar." *Applied Optics* 39, 3604–3611.

Mattsson, J.O., and Bärring, L. (2001). "Heiligenschein and related phenomena in divergent Light." *Applied Optics* 40, 4799–4806.

Medda, F., Nijkamp, P., and Rietveld, P. (2003). "Urban land use for transport systems and city Shapes." *Geographical Analysis* 35, 46–57.

Memory, J.D. (1973). "Kinematics problem for joggers." *American Journal of Physics* 41, 1205–1206.

Mills, E.S. (1970). "Urban density functions." *Urban Studies* 7, 15–20.

Montroll, E.W., and Badger, W.W. (1974). *Introduction to Quantitative Aspects of Social Phenomena*. Gordon and Breach, New York.

Monteith, J.L. (1954). "Refraction and the spider." *Weather* 9, 140–141.

Myrup, L.O. (1969). "A numerical model of the urban heat island." *Journal of Applied Meteorology* 8, 908–918.

Newling, B.E. (1969). "The spatial variation of urban population densities." *Geographical Review* 59, 242–252.

Olfe, D.B., and Lee, R.L. (1971). "Linearized calculations of urban heat island convection effects." *Journal of Atmospheric Sciences* 28, 1374–1388.

Oke, T.R. (1973). "City size and the urban heat island." *Atmospheric Environment* 7, 769–779.

Pearce, C.E.M. (1974). "Locating concentric ring roads in a city." *Transportation Science* 8, 142–168.

Preston-Whyte, R.A. (1970). "A spatial model of an urban heat island." *Journal of Applied Meteorology* 9, 571–573.

Puu, T. (1978). "Towards a theory of optimal roads." *Regional Science and Urban Economics* 8, 203–226.

Rosenau, H. (1983). *The Ideal City: Its Architectural Evolution in Europe*. Routledge and Kegan Paul, London.

Shankland, R.S. (1968). "Rooms for speech and **music**." *Physics Teacher* 6, 443–449.

Shoemaker, E.M. (1983). "Asteroid and comet bombardment of the earth." *Annual Review of Earth and Planetary Sciences* 11, 461–494.

Sitte, C. (1889,1965). *City Planning, According to Artistic Principles*. Random House, New York.

Tan, T. (1966). "Road networks in an expanding circular city." *Operations Research* 14, 607–613.

Theusner, M. (2011). "Photographic observation of a natural fourth-order rainbow." Applied Optics 50, F129–F133.

Vaughan, R.J., et al. (1972.) "Traffic characteristics as a function of the distance to the town center." *Traffic Engineering Control* 14, 224–227.

Vergara, W.C. (1959). *Mathematics in Everyday Things.* Harper, New York.

Walker, J.D. (1975). *The Flying Circus of Physics with Answers.* Wiley, New York.

Walker, J.D. (1976). "Multiple rainbows from single drops of water and other liquids." *American Journal of Physics* 44, 421–433.

Wardrop, J.G. (1952). "Road paper: Some theoretical aspects of road traffic research." *Proceedings of the Institution of Civil Engineers, Part II*, 1, 325–362.

Wardrop, J.G. (1968). "Journey speed and flow in central urban areas." *Traffic Engineering Control* 9, 528–532.

Watson, G.N. (1959). "Note 2871: A quadruple integral." *Mathematical Gazette* 43, 280–283.

Webster, F.V. (1958). "Traffic signal settings." Technical paper 39, Road Research Laboratory, D.S.I.R.

West, G.B. (1999). "The origin of universal scaling laws in biology." *Physica A* 263, 104–113.

Williamson, S.J. (1973). *Fundamentals of Air Pollution.* Addison-Wesley, Reading, MA.

Wolke, R.L. (2000). *What Einstein Told His Barber.* Bantam Doubleday Dell, New York.

Yeo, G.F. (1964). "Traffic delays on a two-lane road." *Biometrika* 51, 11–15.

INDEX